D1535466

Non-Linear Structures

Matrix Methods of Analysis and Design by Computers

Non-Linear Structures

Matrix Methods of Analysis and
Design by Computers

K. I. Majid, B.Sc., Ph.D., M.I.C.E.,
M.I. Struct. E.

(Professor of Civil Engineering, The University
of Aston, Birmingham)

WILEY – INTERSCIENCE

A Division of John Wiley & Sons, Inc.,

New York — London — Sydney — Toronto

Published in the USA by
Wiley-Interscience Division,
John Wiley & Sons, Inc.,
605, Third Avenue, New York, N.Y. 10016

Library of Congress Catalog Card Number: 76-39367

First published 1972
© Butterworth & Co. (Publishers) Ltd.,
ISBN 0 408 70251 6 (Butterworth)
ISBN 047156535-0 (Wiley)

Printed in Hungary

Preface

A characteristic feature of twentieth century engineering is tall and slender structures in which non-linear and secondary effects are preponderant. For this reason a great deal of attention has been given to studying these effects and the classical theory of columns has been extended to deal with complete structures. In particular, the combined effects of stability and plasticity are of paramount importance in the field of structural engineering, so much so that both plasticity and stability effects in frames have now been introduced into degree courses. In many universities, postgraduate M.Sc. and Diploma courses in non-linear structures are also well established. Furthermore a vast number of research workers, having exhausted the linear aspects of structures, have now moved on to the study of non-linearity. For these people a comprehensive book covering various aspects of non-linearity is of value.

The effect of non-linearity becomes more apparent with larger structures where recent advances in matrix methods of structural analysis become necessary and the aim of this book is to outline the use of modern methods suitable for computers. In using matrix methods the problems of non-linearity follow naturally from the linear ones and are much easier to understand.

The text deals with the problems of both analysis and design. Structures with non-linear behaviour, whether elastic or elastic-plastic have many new features and side effects necessitating a new approach to their design. The science of operational research characterised in optimisation by linear and non-linear programming is becoming more useful in all walks of life. In structural engineering it is giving, for the first time, a scientific basis to the design of structures and the main aspects of recent work carried out on this subject are fully covered in this book.

In the first chapter a brief introduction to the behaviour of structures in general is given with reference to the linear elastic and simple plastic methods of structural analysis. Chapter 2 is devoted to linear matrix methods, both force and displacement, and the matrices developed here are used in later chapters. There is sufficient information in this chapter to enable the reader to understand the matrix methods of structural analysis, but a knowledge of basic Matrix Algebra is assumed.

Chapter 3 studies the stability of an individual member with various end conditions. It also derives the stability functions used in Matrix force and Matrix displacement methods. Chapter 4 covers the elastic stability of complete frames, whilst Chapter 5 deals with the elastic instability of frames. In the latter chapter, attention is given to instability as an eigenvalue problem and the vanishing characteristic of the determinant of the stiffness matrix is presented to study the state of frame instability. The general critical state of frames is formulated and, from this, the other relevant elastic critical modes will be deduced.

Chapter 6 presents the elastic-plastic analysis of frames, taking the non-linearity due to stability and plasticity into consideration. The method of evaluating the failure load of structures by following the sequence of hinge fromation is developed in detail. This is followed in Chapter 7 by a number of approximate methods for the evaluation of the failure load of frames without following the sequence of hinge formation.

The last three chapters are devoted to the design of structures and the non-linear aspects of design problems. These cover: The theories on non-linear elastic-plastic analyses; recent developments in mathematical optimisation to produce minimum weight design methods for structures where strength as well as deflections form important aspects of design; and a summary of linear programming and the simplex method. A description of non-linear programming by piecewise linearisation is included in Chapter 10. Using Matrix force method, detailed procedures are given for the minimum weight design of statically determinate and indeterminate structures by non-linear programming.

Exercises are given at the end of each chapter and, where available, experimental verification of the theoretical work is also provided. A number of computer programmes are discussed, each programme being presented by giving its procedure together with a flow diagram.

The author acknowledges the help of present and former colleagues at the Universities of Manchester and Aston who have worked with him on the subject matter of this book. He wishes to thank Professor M. R. Horne as the author's first teacher on the subject of non-linear analysis of structures. In particular the author is grateful to Mr. Alan Jennings, senior lecturer at the Queens University of Belfast, who collaborated on many of the ideas in the early part of this book. Thanks are also due to Professor M. Holmes for his support in the writing of this book, reading the text and making valuable suggestions, and to Dr. D. Anderson who carried out research with the author into a number of topics.

The author particularly wishes to thank Dr. H. W. Sinclair-Jones who was kind enough to study the text, resolve the examples and check the diagrams.

K. I. MAJID

Contents

tation of a member — Sway of a member — A state of general sway of a member with hinges — The stability functions used in the matrix displacement method — The stability functions used in matrix force method — Relationships between the stability functions — General deformation of a member with plastic hinges — Effect of lateral loading and gusset plates — A computer routine for the evaluation of the stability functions — Exercises.

culation of failure loads–Approximate method 1 — Assumptions — Analysis of columns — Analysis of frames — Rankine's formula–Approximate method 2 — Experimental evidence — Exercises.

Notations

Symbols are defined when they appear for the first time in the text. Each one is redefined when its meaning changes. Some symbols are used throughout the text with the following meaning.

a	Constant
b	Breadth of a section
c	Stability function
d	Depth of section
h	Height
k	EI/L
I_p, m_p, n_p	Direction cosines for P axis
m, n, o	Stability functions
p	Force in a member
q, r	Constant ratios
s	Stability functions
t	Thickness
u, v	Member displacements parallel to P and Q axes respectively.
x, y	Joint displacements parallel to X and Y axes respectively
z	Objective function
A	Member area
B	Breadth
D	Depth
E	Modulus of elasticity
H	Force parallel to X axis
I	Second moment of area
K	Stiffness

L	Span, length
M	Moment
M_p	Fully plastic moment
P	Force
P, Q, R	Local member coordinates or axes
S	Shear force
V	Force parallel to Y axis
W	Force
X, Y, Z	Overall reference coordinates
Z	Section modulus

Matrices, Vectors and Elements

a	diagonal matrix of reciprocal of member areas
a_{11}, a_{12}	Submatrices of overall stiffness matrix K corresponding to L_1
A	Displacement transformation matrix
A'	Transpose of A
a	Element EA/L
B	Load transformation matrix
B_b, B_r	Submatrices of B corresponding respectively to the basic statically determinate structure and redundant forces.
b	Element $12\,EI\phi_5/L^5$
D	Value of determinant
d	Element $-6EI\phi_2/L^2$
e	Element $4EI\phi_3/L$
f	Element $2EI\phi_4/L$
f	Member flexibility matrix
F	Overall flexibility matrix of structure
I	Unit matrix
k	Member stiffness matrix
K	Overall stiffness matrix of structure
L	External load matrix or vector
L_1, L_2	Submatrices of L
L_b	External force matrix or vector applied to basic statically determinate structure
L_r	Redundant force matrix or vector
O	Null matrix
P	Member force vector
X	Joint displacement matrix or vector
X_1, X_2	Subvectors of X corresponding to L_1, L_2

X_b, X_r	Displacement vectors corresponding to L_b and L_r
Z	Member displacement or distortion vector
Δ	Upper bound displacements vector
σ	Vector of permissible stresses

Greek Symbols

α	Shape factor
γ	Angle of pitch
Δ, δ	Deflections
σ	Stress
θ	Rotation
λ	Load factor
ϕ	Δ/L
ϕ_1 to ϕ_7	Stability functions
ϱ	P/P_E ratio of axial load to Euler load for pin-jointed member
$\propto$	Angle

CHAPTER 1

Introduction

1.1. BEHAVIOUR OF IDEALISED STRUCTURES

The stress strain relationship for specimens of structural materials, such as mild steel, has the typical form shown in Figure 1.1. It is almost exactly linear in the 'elastic range' until the 'upper yield'

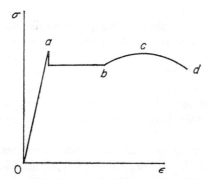

Figure 1.1. Stress-strain diagram for mild steel

stress is reached at *a*. The stress then drops abruptly to the 'lower yield' stress and the strain then increases at constant stress up to the point *b*, this behaviour being termed 'purely plastic' flow. Beyond point *b* a further increase of stress is required to produce an increase in strain, and the material is said to be in the 'strain-hardening'

1

range. Eventually a maximum stress is reached at *c*, beyond which increases in strain occur with decrease of stress until rupture occurs at *d*. The slope of the linear part *Oa* gives Young's modulus and the ratio of the upper yield stress to the lower is in the order of 1·25 for steel.

In analysing structures, various idealisations are made to render the approach manageable. Although the upper yield phenomenon is a real one, it vanishes on cold-working and is often not exhibited by the material of rolled steel sections. The first idealisation is thus

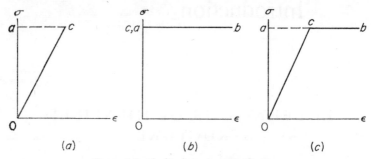

Figure 1.2. Idealised stress-strain diagrams
(a) elastic
(b) rigid-plastic
(c) elastic-plastic

to ignore this upper yield and this approach has been adopted by nearly all those who have dealt with problems of plasticity and instability.

The effect of strain hardening on the carrying capacity is considerable, especially with small structures. For tall structures, however the effect of instability is far greater than of strain hardening. In these structures it is reasonable therefore, to neglect this effect also.

Apart from the above idealisations which are nearly universal, various analytical theories require further simplification. The elastic theory, for instance assumes that the safety factor used, safeguards against yielding and thus restricts itself to the portion *Oc* of the stress-strain diagram, Figure 1.2(a). The 'rigid plastic' theory, on the other hand, is based on the idealisation that the stress-strain relation *Ocb* of Figure 1.2(b) reproduces closely the stress-strain relation for the material concerned. This shows no strain up to point *c* followed by a pure plastic flow *cb*. Finally the elastic-plastic

2

theory assumes that the stress-strain relation is that given by *Ocb* in Figure 1.2(c). This is an improvement on the plastic-theory since the latter neglects the elastic strain energy represented by the area *Oac*.

1.2. VARIATION OF GENERALISED DEFORMATION OF STRUCTURES WITH THE LOAD PARAMETER

The calculated values of the lateral deformations of the members of a rigid structure measured orthogonally to their original directions are determined by the assumptions made concerning the stress-strain relationship of the material of the structure. They also depend upon the manner in which the external loads are applied as well as the forces that are considered to be significant when deriving the equilibrium equations.

For instance, if the material is assumed to be perfectly elastic and the applied loads are axial to the members of the structure, then no displacement takes place in the members until the load factor reaches a critical value resulting in large deformations at this load. This is shown by curve *OAC* in Figure 1.3. The behaviour of a perfect elastic pin ended strut is a well known example. The value λ_c is called the elastic critical load factor of the structure. This is equivalent to the Euler load of a single member.

The same structure maybe subjected to a set of 'non-axial' loads. That is to say the loads are applied to the members between the joints and give rise to transverse bending as well as axial loads in the various members. If the effect of the resulting axial loads on the equilibrium conditions are neglected, the resulting lateral displacements, due to the bending effects alone, vary linearly with the load parameter. This state of deformation is represented by the straight line *OB* in Figure 1.3. An analysis of the structure in this manner is known as the 'Linear elastic analysis'. A point such as *G* on the line *OB* can be obtained by any of the well-known linear elastic approaches such as the slope-deflection method.

On the other hand, the 'simple plastic' analysis assumes a 'rigid-plastic' stress-strain relationship as shown in Figure 1.2(b). Once again neglecting the axial loads in the members the load deformation characteristic of the structure follows the vertical line *OD*,

Figure 1.3, until suddenly 'collapse' takes place and deformation increases indefinitely as shown by the horizontal line *DE*.

Apart from neglecting any elastic deformation of the structure, prior to collapse, the simple plastic theory involves a number of other assumptions. For instance it assumes that an increment of the bending moment at a section always causes an increment of curvature of the same sign and that the magnitude of the curvature always tends to become abruptly large when the bending moment reaches the 'fully plastic' moment of the cross-section.

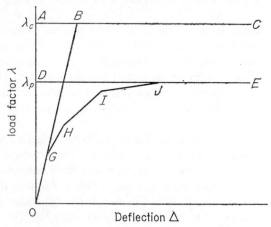

Figure 1.3. Linearised load deformation of structures

The theory also assumes that whenever the fully plastic moment is attained at any cross-section, a 'plastic hinge' is formed there which can undergo rotation of any magnitude so long as the bending moment remains constant at the full plastic value. Once a hinge is introduced at a section, it is assumed that it will continue to rotate in the same direction.

According to the 'uniqueness' theorem of the simple plastic approach a structure collapses when three requirements are satisfied. These are the mechanism requirement, that is to say a sufficient number of plastic hinges should develop to turn the structure, or part of it, into a mechanism; the equilibrium requirement i.e. the internal bending moments should be in equilibrium with the applied loads; and the yield requirement, namely the bending moment should nowhere in the structure be greater than the fully plastic moment of the cross section.

4

Frames with more realistic elastic-plastic material, Figure 1.2(c), however, follow the piece-wise linear elastic-plastic curve shown as *OGHIJE* in Figure 1.3. The 'piece-wise linear elastic-plastic' theory assumes that the load-deflection curve of a structure follows the elastic line *OG* until a plastic hinge is formed at a cross-section where the bending moment is the highest in the structure and equal to the fully plastic moment of the section. From then onwards the deflections increase at a faster rate as shown by *GH* in the figure until a second plastic hinge develops in the structure. This process continues until the applied load factor reaches that given by the rigid-plastic theory and collapse takes place. The portions *OG*, *GH*, *HI* etc., are considered to be linear, while the overall non-linearity in the load-deflection relationship is purely due to the development of material plasticity at discrete sections. The piece-wise linear elastic-plastic theory involves exactly the same assumptions as the rigid plastic theory except that the pre-collapse deflections of the structures are not neglected.

1.3. LOAD-DISPLACEMENT DIAGRAMS OF A PORTAL

As an example let us consider the fixed base portal frame loaded as shown in Figure 1.4. and analyse it using the three different approaches discussed in the previous section. This portal will be analysed in subsequent chapters using various non-linear theories. Its dimensions and cross-sections are selected in a way that makes the frame very flexible so that the sets of non-linear curves may be readily distinguished.

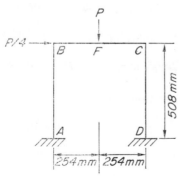

Figure 1.4. Fixed-base portal frame

The frame has span L of 508 mm and the span to height ratio is unity. The cross-section of the members is $b \times d$ which is uniform throughout with $b = 12 \cdot 7$ mm and $d = 3 \cdot 175$ mm. The modulus of elasticity E is 207 kN/mm^2 and the fully plastic moment M_P is $9 \cdot 27$ kNmm. For the sake of simplicity it is assumed that a fully plastic hinge develops at a section as soon as the extreme fibre stress reaches the yield stress. In other words the 'shape factor' of the section is unity. For the linear elastic analysis denoting the stiffness EI/L of the cross section by k, the rotation at a joint by θ, and the horizontal sway by Δ, then under the loads shown the bending moments M at the ends of the members can be calculated using the slope-deflection equations, i.e.

$$
\begin{aligned}
M_{AB} &= 2k\theta_B - 6k\Delta/L \\
M_{BA} &= 4k\theta_B - 6k\Delta/L \\
M_{BC} &= 4k\theta_B + 2k\theta_C - PL/8 \\
M_{CB} &= 4k\theta_C + 2k\theta_B + PL/8 \\
M_{CD} &= 4k\theta_C - 6k\Delta/L \\
M_{DC} &= 2k\theta_C - 6k\Delta/L
\end{aligned}
\tag{1.1}
$$

For the equilibrium of joint B

$$\Sigma M_B = M_{BA} + M_{BC} = 0$$

and from equations (1.1), it follows that

$$8k\theta_B + 2k\theta_C - 6k\Delta/L = PL/8 \tag{1.2}$$

Similarly for joint equilibrium at C

$$2k\theta_B + 8k\theta_C - 6k\Delta/L = -PL/8 \tag{1.3}$$

Finally the shear equation gives

$$PL/4 + M_{AB} + M_{BA} + M_{CD} + M_{DC} = 0$$

which, upon substitution from the slope-deflection equations (1.1), yields:

$$PL/4 + 6k\theta_B + 6k\theta_C - 24k\Delta/L = 0 \tag{1.4}$$

6

Solving equations (1.2) through (1.4) gives:

$$\Delta = \frac{5PL^2}{336k}$$

$$\theta_B = \frac{5PL}{168k}$$

(1.5)

$$\theta_C = -\frac{PL}{84k}$$

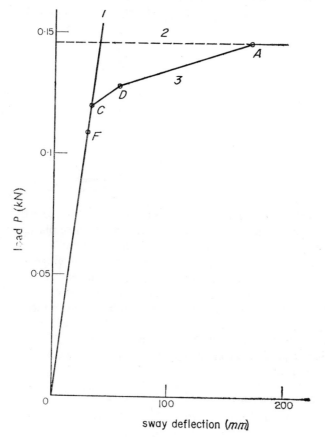

Figure 1.5. Load displacement graphs of portal
Curve (1) Linear elastic
Curve (2) Rigid plastic
Curve (3) Piece-wise linear elastic-plastic

7

These equations show that the horizontal sway and the joint rotations are linearly related to the applied load P. In Figure 1.5 the values of Δ are plotted against P as shown by the straight line (1). The slope-deflection equations (1.1) can now be used to calculate the bending moment distribution throughout the frame. This is shown in Figure 1.6 in terms of P and to three figures accuracy.

The analysis of the portal by the plastic theory can be performed in accordance with the 'kinematic theorem' that states: 'Of all the mechanisms formed by assuming plastic hinge positions, the correct mechanism for collapse is that which requires the minimum load'.

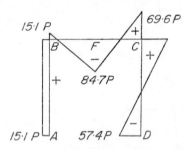

Figure 1.6. Linear elastic bending moments

The portal frame is statically indeterminate with three redundancies, hence it becomes statically determinate if three plastic hinges form in appropriate places and will collapse with a mechanism of four hinges. In this frame there are five possible hinge locations at A, B, C, D and F and therefore $5-3 = 2$ independent mechanisms can develop in the frame. These are the 'sway mechanism' with hinges forming at A, B, C and D and the 'beam mechanism' with hinges forming at B, F, and C as shown in Figure 1.7(a) and (b). Each one of these mechanisms corresponds to an equilibrium equation which can be derived using the virtual work method, thus

Sway mechanism: $\qquad PL \, \delta\theta/4. = 4M_P \, \delta\theta$ $\qquad$ (1.6)

Beam mechanism: $\qquad PL \, \delta\theta/2. = 4M_P \, \delta\theta$ $\qquad$ (1.7)

where $\delta\theta$ is an angular movement as shown in Figure 1.7. A 'combined mechanism' can be obtained from the above two. This is shown in Figure 1.7(c) in which the hinge at B is eliminated. The

8

virtual work equation for this new mechanism is

$$PL\,\delta\theta/2 + PL\,\delta\theta/4 = 6M_p\,\delta\theta \qquad (1.8)$$

From equations (1.6) through (1.8) it follows that according to the kinematic theorem the collapse load of the frame is $P = 8M_p/L$ given by both equations (1.7) and (1.8). Using the appropriate values of M_p and L given earlier the frame collapses at a load of 0·146 kN. The dotted line (2) in Figure 1.5 has this ordinate.

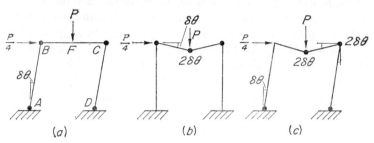

Figure 1.7. Rigid plastic mechanisms
(a) Sway mechanism
(b) Beam mechanism
(c) Combined mechanism

It is of interest to note, in this example, that the condition for collapse both with beam mechanism and combined mechanism is obtained at the same value of $P = 8M_p/L$. Such a state is termed 'overcollapse' and the frame will develop five hinges at A, B, F, C and D. In reality, because of some imperfection, the frame will collapse either with beam mechanism or with combined mechanism. Indeed, theoretically, if the side load is reduced slightly the frame will collapse with beam mechanism while a small increase of the side load leads to collapse with combined mechanism.

The piece-wise linear elastic-plastic analysis of the frame is somewhat difficult. Perhaps a convenient approach is as follows. From the linear elastic bending moment diagram, shown in Figure 1.6, it is noticed that the maximum bending moment occurs at F with a value of 84·7P. Naturally, if the value of P is increased, the first plastic hinge would develop at this point and the value of the bending moment would be equal to the fully plastic hinge moment M_p of the section. Thus with $M_p = 9·27 \times 10^{-3}$ kNm, the load which causes the development of this first plastic hinge at F is

9

$P = 0.109$ kN. The elastic sway deflection corresponding to this state can now be evaluated from the first of equations (1.5).

Once the value of P exceeds 0.109 kN the plastic hinge developed at F begins to rotate. The frame is now more flexible and the overall deflections increase faster than before as shown by the straight line GH in Figure 1.3. Point H on this line represents the state where a second plastic hinge is about to develop at joint C in the frame, where the elastic bending moment is the highest. In order to locate point H, the slope deflection method can, once again, be used but this time making use of the fact that the bending moment at F would have a constant value of M_p with the discontinuous rotations θ_{FC} and θ_{FB} being unequal. The bending moment at C would also be equal to the fully plastic moment, except for the fact that since this value has just been attained without, as yet, any hinge rotation, the condition of continuity still exists at C with $\theta_{CF} = \theta_{CD} = \theta_C$. The slope deflection equations will therefore become

$$
\begin{aligned}
M_{AB} &= 2k\theta_B - 6k\,\Delta/L \\
M_{BA} &= 4k\theta_B - 6k\,\Delta/L \\
M_{BF} &= 8k\theta_B + 4k\theta_{FB} - 24k\,\delta/L \\
M_{FB} &= -M_p = 8k\theta_{FB} + 4k\theta_B - 24k\,\delta/L \\
M_{FC} &= M_p = 8k\theta_{FC} + 4k\theta_C + 24k\,\delta/L \\
M_{CF} &= M_p = 8k\theta_C + 4k\theta_{FC} + 24k\,\delta/L \\
M_{CD} &= -M_p = 4k\theta_C - 6k\,\Delta/L \\
M_{DC} &= 2k\theta_C - 6k\,\Delta/L
\end{aligned}
\tag{1.9}
$$

where δ is the vertical deflection at F. It is noticed that in the slope deflection equations point F is treated as a structural joint in the frame and thus the stiffness of the two portions BF and FC is $2EI/L$.

The value of $k\theta_{FB}$ can be eliminated from the third of the equations by using the fourth since the latter has a known left-hand side, thus

$$
M_{BF} = 6k\theta_B - 0.5M_p - 12k\,\delta/L
$$

Similarly using the 5th, 6th and 7th by eliminating $k\theta_{FC}$ and $k\theta_C$ it follows that

$$
k\,\Delta/L = \frac{2M_p}{9} - \frac{4k\,\delta}{3L}
\tag{1.10}
$$

10

For the joint equilibrium at B, $\Sigma M_B = 0$, hence

$$10k\theta_B = \frac{6k\,\Delta}{L} + \frac{12\,k\delta}{L} + 0\cdot 5M_p \qquad (1.11)$$

The horizontal sway equation is

$$\tfrac{1}{4}\,PL + M_{AB} + M_{BA} + M_{CD} + M_{DC} = 0,$$

which upon substitution from equation (1.9) yields

$$\frac{1}{4}\,PL + 6k\theta_B - 15k\,\frac{\Delta}{L} - \frac{3M_p}{2} = 0 \qquad (1.12)$$

A second sway equation can be obtained using virtual work equation for vertical loading, i.e.

$$\tfrac{1}{2}\,PL + M_{BF} + M_{FB} - M_{CF} - M_{FC} = 0$$

which gives

$$\frac{1}{2}\,PL - \frac{7}{2}\,M_p + 6k\theta_B - 12k\,\frac{\delta}{L} = 0 \qquad (1.13)$$

Equations (1.10) through (1.13) can now be solved for the four unknowns θ_B, Δ, δ and the load P. Using the known value of $9\cdot27 \times 10^{-3}$ kNm for M_p in these equations, the condition at which the second hinge develops at C is found to be when $P = 0\cdot120$ kN and Δ is 33 mm.

The above procedure can be used to complete the piece-wise linear elastic-plastic graph. It is noticed from Figure 1.6 that the third plastic hinge will develop at D and the last one at A.

Once again, when considering the condition for the formation of the third hinge at D, as represented by point I in Figure 1.3, the bending moment at D would be equal to the fully plastic moment of the section while due to continuity θ_D will be zero. Because of discontinuity at F and C, on the other hand $\theta_{FC} \neq \theta_{FB}$ and $\theta_{CF} \neq \theta_{CD}$. The details of the analysis for this and for the last case when a hinge develops at A are left as an exercise to the reader. The third hinge at D will form when the applied load P is $0\cdot128$ kN and the corresponding lateral sway is 57 mm. Finally the formation of the last hinge at A takes place when P reaches the rigid plastic collapse load of $0\cdot146$ kN with a sway of 171 mm. The piece-wise linear elastic-plastic graph is shown as curve (3)

11

in Figure 1.5. The letters on this curve refer to the position on the frame where the hinges are developed.

The foregoing analysis appears to be rather lengthy but is in fact one of the quickest approaches. It has the advantage that at every stage all the information concerning the deflections and bending moment distribution as well as the hinge rotations are readily available. The lengthy appearance of the process indicates why so little is written about the non-linear behaviour of structures.

1.4. THE EFFECT OF AXIAL LOAD ON THE BEHAVIOUR OF IDEALISED STRUCTURES

In the preceding sections three different theories were summarised for the analysis of ideal structures. Apart from the idealisations stated concerning the material properties, all three theories are based on two further assumptions. First it is assumed that during the loading process of a structure the change in the actual shape and dimensions of the structure is immaterial. Thus they all depend on the small deflection theory which permits one, when considering a structural deformation θ, to make the approximation that $\cos \theta = 1$ and $\sin \theta = \tan \theta = \theta$. In the example of the portal frame it was calculated that at the formation of the fourth and last hinge at A, the sidesway of the frame was 171 mm and thus the columns of the frame were inclined some 20° to the vertical. Naturally the use of small deflection theory with such a deformation is bound to lead to gross errors. However, with more realistic frames, the overall deformation is not so large and the small deflection theory may be sufficiently accurate.

The second assumption underlying all the three approaches is that because the deformations are small, the effect of axial loads in the members can be neglected when deriving the equilibrium equations for a structure. For a general member such as that shown in Figure 1.8, it is assumed that for the equilibrium of the member

$$M_A + M_B = SL \qquad (1.14)$$

where S is the shear force in the member.

It is seen from the figure that when either the lateral deformation of the member or the axial load P in the member or both are large, equation (1.14) becomes inaccurate and should be altered to read

$$M_A + M_B + P\Delta = SL \qquad (1.15)$$

A primary cause of non-linearity in structures is due to this effect of P on the equilibrium of the structure. At a distance x from end A of the member, see Figure 1.8, the axial load in the member causes an additional bending moment of magnitude $P(Y_1 + Y_2)$. This additional moment is due to two different types of deformation. The

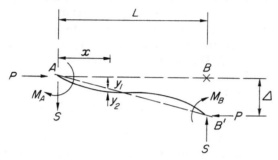

Figure 1.8. General deformation of a member
Y_1 is due to joint translation
Y_2 is due to member curvature

amount PY_1 is due to the translation of joint B to B' by Δ relative to A, with $Y_1 = \Delta x/L$, while PY_2 is due to the actual curvature in the member resulting from the presence of end moments M_A and M_B.

A difficulty in the non-linear structural analysis arises from the fact that the axial forces in the members are themselves unknown and cannot be included in the derivation of the equilibrium equations. This problem will be studied extensively in the subsequent chapters. In the meantime it is advantageous to discuss briefly the effects of the member axial load upon the behaviour of structures with particular reference to the three theories presented earlier in this chapter.

Consider first, frames subject to applied loads that are all axial to the members. These axial loads do not affect the load deformation characteristics of ideally perfect structures until the elastic critical load factor λ_c is reached. At this load factor, once the structure begins to deform, the axial loads start to play a part in the equilibrium of the members, resulting in modifying the straight line

13

AC of Figure 1.3 to the drooping curve *AC'*, shown in Figure 1.9. That is to say to maintain equilibrium, as deflections increase, the externally applied loads have to be reduced in magnitude.

Similarly, for non-axially loaded elastic structures, taking the effect of the resulting axial loads into consideration modifies the lateral displacement in a non-linear manner as indicated by curve *OC''* in Figure 1.9. Flexible triangular structures, where primary bending moments in the members are small may behave in this manner. As the compressive axial loads in the members increase, the stiffness of these members is reduced, leading to a reduction in the overall stiffness of the structure. This causes the elastic failure of the structure at a load factor λ_{EF} which is below the elastic critical load. The elastic failure load is often confused with the elastic critical load. The distinction between the two is in the fact that the former is the maximum load the structure would carry under any set of loads provided the material remains entirely elastic. The elastic critical load, however is a state of bifurcation of equilibrium such that, under a set of idealised loads, the structure may either remain undisturbed or assume a mode of deformation.

Consideration of the axial load also alters the rigid plastic behaviour of structures by modifying the straight line *DE* to the drooping curve *DE'*. Far more important than this, however, is the fact that consideration of axial loads undermines the basic fundamentals of the rigid plastic theory itself. The axial loads violate the three requirements of equilibrium, yield and mechanism stated earlier under the uniqueness theorem. Firstly if a mechanism is assumed at collapse then by considering the axial loads the equilibrium conditions, as identified by the virtual work equations, are different from those suggested by the rigid plastic theory. The virtual work equations will now contain second order $\delta\theta^2$ terms, where $\delta\theta$ is the incremental deformation given in equations (1.6) through (1.8). Equation (1.8) for the combined mechanism of the portal for instance changes to become

$$(PL\ \delta\theta/2 + PL\ \delta\theta/4) + LR_A\ \delta\theta^2 + LR_D\ \delta\theta^2 + LH\ \delta\theta^2 = 6M_P\ \delta\theta \quad (1.16)$$

where the terms between the brackets are those used in equation (1.8). The remaining terms on the left-hand side of equation (1.16) are due to the axial load effects. Here R_A, R_D and H are the vertical reactions at *A* and *D* and the horizontal thrust respectively. Derivation of equation (1.16) will be given in a later chapter.

Furthermore, introducing the effect of axial loads also violates

14

the yield criterion by making the bending moments at certain sections, not necessarily at joints or under point loads, higher than the fully plastic moments of these cross-sections.

The worst effect of the axial loads, however, is to make collapse take place before the formation of a mechanism and at load factor λ_F which is below that predicted by the rigid plastic theory. During the loading process, the resulting axial loads reduce the overall

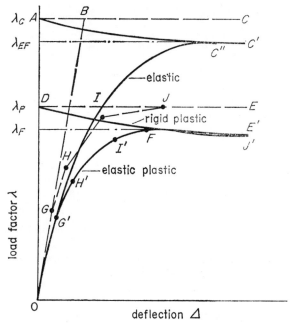

Figure 1.9. Non-linear load deformations of structures

stiffness of the structure continuously and modify the piece-wise linear elastic-plastic response to that shown by $OG'H'I'FJ'$ in Figure 1.9, where F is a point with the highest ordinate on the curve and indicates the pre-mechanism state of failure.

Finally, the axial loads may alter the mode of deformation completely in such a way that the sequence of hinge formation would be entirely different from those predicted by the piece-wise linear elastic-plastic theory. To demonstrate this the two storey frame of Figure 1.10, originally analysed by Horne[1], is considered. The load W_p is such that rigid-plastic collapse of the frame occurs

15

when the load factor λ is unity. The members of the frame are made out of two parallel plates so that the shape factor is unity. The depth between the plates is $2r$ and the slenderness ratio L/r of the members is 100. The modulus of elasticity is 207 kN/mm² and the yield stress is 0·248 kN/mm². For many reasons this frame is ideal for a rigid-plastic or piece-wise linear elastic-plastic analysis. The results of this analysis is shown in Figure 1.10(b), where the

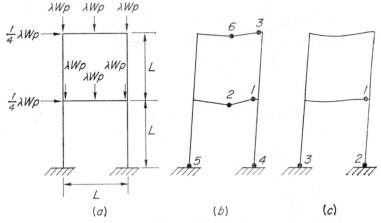

Figure 1.10. Sequence of hinge formation for a two-storey rigid frame
(a) Frame and loading
(b) Rigid-plastic mechanism with sequence of hinge formation by piece-wise linear elastic-plastic analysis
(c) Sequence of hinge formation by non-linear elastic-plastic analysis

sequence of hinge formation is shown by the numbers next to each hinge. Collapse takes place with a combined mechanism with six hinges.

A non-linear elastic-plastic analysis, taking the effect of axial loads into consideration, was carried out for the same frame and the results of this are shown in Figure 1.10(c), where it is seen that the frame fails with only three hinges, much before the formation of a mechanism. The sequence of these three hinges is entirely different from the first three hinges predicted by the piece-wise linear elastic-plastic analysis. At collapse, which takes place at $\lambda = 0·777$, the top storey of the frame remains elastic and the overall sidesway of the frame is far less than that given by the piece-wise linear analysis.

16

1.5. THE REAL BEHAVIOUR
OF STRUCTURES

So far the two causes of non-linearity in structures have been attributed to the effects of axial loads in the members and of plasticity of the material. In real structures there are numerous other factors that play an important part in aggravating the non-linearity of structures. A few of those that are easily conceived are the effect of the shape factor not being unity, the strain hardening effect mentioned earlier, the effect of uneven and non-linear behaviour of the foundations, the influence of the non-linear behaviour of cladding, infills and plates on the frames etc. Reinforced concrete structures are becoming more common, particularly in the construction of tall buildings and the stress-strain or moment curvature relationships of this material are highly non-linear, involving discontinueties. An important cause of non-linearity in reinforced concrete structures is due to the cracking of concrete. It is not so easy to define the fully plastic moment of reinforced concrete sections because at an early stage of loading, before the formation of a plastic hinge, the concrete begins to crack resulting in a reduction in the moment of resistance of the section.

To give a full treatment to every aspect of non-linearity is beyond the scope of this book; therefore, only the main aspects of non-linearity will be covered. Attention will be paid both to the analysis aspects and to the design of structures, where the problem of non-linearity is more complex. Because of the complex nature of the problem, it is realised that the use of computers and modern techniques, such as matrix methods, are unavoidable. In the next chapter, therefore, the matrix methods of structural analysis will be presented, and these will be used throughout the succeeding chapters.

The plastic theory of structures is perhaps the simplest of the theories currently studied by undergraduates and applied, in a limited way, by practising engineers. Many aspects of the plastic theory are used throughout this book, both when dealing with the elastic-plastic analysis and with structural design. The remainder of this chapter will cover this theory.

1.6. THE RIGID-PLASTIC THEORY

It was pointed out earlier that the rigid-plastic theory neglects the pre-collapse strains developed in the material of the structure. At the instant of collapse, it assumes that the geometry of the structure is unaltered. Nevertheless it suggests that, at collapse, the externally applied loads are sufficiently high to cause the yielding of the material at certain sections so that the structure behaves like a mechanism. The structure at this stage is statically determinate and hence the equations of static equilibrium are utilised to calculate the collapse load.

In spite of these assumptions, the rigid plastic theory can be usefully employed to calculate the carrying capacity of many structures, particularly in the case of continuous beams and simple frames. The development of the rigid-plastic theory contributed to a better understanding of the state of collapse. In particular, it was via the plastic theory that the ultimate load theory became a method for the design of structures. Unfortunately because of this the concept of the ultimate load carrying capacity of a structure is often confused with that given by the plastic theory. It is advantageous therefore, to present the plastic theory first and then proceed with the study of the ultimate failure load theory in detail. This procedure is necessary since both theories involve many similar concepts such as the proportional method of loading, the plastic hinge formation and others. It is convenient to introduce these concepts with the simpler of the two theories, namely the plastic theory.

1.7. THE FULLY PLASTIC HINGE MOMENT

Consider a cross-section of a structural member that is subject to a moment which is increasing gradually. While the cross-section is elastic the stress distribution across the section is linear as shown in Figure 1.11(a). At a certain value of the bending moment the extreme outer fibres begin to yield and thereafter remain at a constant stress σ_y.

Increasing the moment further causes the spread of yielding to the inner fibres and the stress distribution across the section takes the form shown in Figure 1.11(c). At some stage the bending moment reaches a certain value M_p known as the fully plastic moment, and the entire cross-section yields. The diagram for stress distribution takes the form shown in Figure 1.11(d). From then onwards the section becomes unable to sustain an increasing moment. At this constant moment and fibre stress σ_y, the section begins to deform plastically allowing the development of large strains across the section and large deformation of the structural member. It is said that the section has developed a plastic hinge.

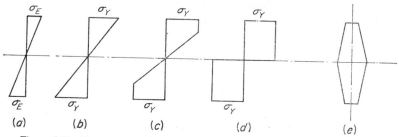

Figure 1.11. Cross-section of structural member subject to increasing moments (a) Elastic; (b) first yield; (c) spread of yield; (d) fully plastic; (e) section with a vertical axis of symmetry

For a symmetrical section the plastic moment of resistance, which is also equal to M_p, can be calculated from a knowledge of the dimensions of the section and the value of the yield stress of the material. In the case of the section shown in Figure 1.11(e), the stresses above and below the horizontal axis of symmetry add up to a pair of forces each being equal to $A\sigma_y/2$. The moment $\frac{1}{2}A\sigma_y\times 2d$ of these equal and opposite forces balances the fully plastic moment acting on the section, thus

$$M_p = A\sigma_y d$$

For instance for a rectangular section of width B and depth D, Figure 1.12, the resultant force acting on each half is $\sigma_y BD/2$. These forces act at the centroid of each portion and thus are $D/2$ apart. Hence for this section

$$M_p = \tfrac{1}{2}\sigma_y BD\times D/2 = BD^2\sigma_y/4 \qquad (1.17)$$

It should be pointed out that, unlike an elastic section, where the centroidal axis is the neutral axis, a plastic section has its equal area axis as the neutral axis. This is because the force acting on an element $\mathrm{d}A$ of an area is $\sigma_y \mathrm{d}A$ and the moments of these forces about the neutral axis are zero, thus,

$$\int \sigma_y \, \mathrm{d}A = \sigma_y \int \mathrm{d}A = 0$$
$$\therefore \int \mathrm{d}A = 0$$

This means that the area above the neutral axis is equal to that below it.

As an example consider the I section shown in Figure 1.12(b). The resultant of the forces acting on each flange is $Bt_2\sigma_y$. These are $D-t_2$ apart. On the other hand each half of the web is subject to a

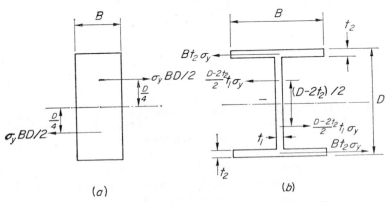

Figure 1.12. Forces acting on sections that are yielding fully
(a) rectangular section; (b) I section

force $(D-2t_2)\sigma_y t_1/2$ and the two forces are $(D-2t_2)/2$ apart. Thus for the I section

$$M_p = [Bt_2(D-t_2) + \tfrac{1}{4} t_1(D-2t_2)^2]\sigma_y$$

1.8. THE SHAPE FACTOR

It has been shown that the fully plastic moment of a section is the product of the yield stress and a quantity which is dependent on the dimensions of the section, i.e.

$$M_p = Z_p\sigma_y, \tag{1.18}$$

where Z_p is known as the plastic section modulus of the section.

20

For a rectangular section for instance, equation (1.17) gives Z_p as $BD^2/4$.

The plastic section modulus is analogous to the elastic section modulus Z. In fact the ratio Z_p/Z is known as the shape factors $\propto$ of the section. For a rectangular section

$$\propto = \frac{Z_p}{Z} = \frac{BD^2}{4} \div \frac{BD^2}{6} = 1\cdot5$$

similarly for an I section, the shape factor $\propto = 1\cdot15$. For a pair of parallel plates the shape factor is unity.

When the outer fibres of a section first yield the elastic bending moment at first yield M_y is equal to $Z\sigma_y$, thus

$$\propto = \frac{Z_p}{Z} = \frac{\sigma_y Z_p}{\sigma_y Z} = \frac{M_p}{M_y} \qquad (1.19)$$

It follows that the shape factor is the ratio between the fully plastic moment of the section and the moment at first yield.

1.9. THE LOAD FACTOR

Consider a simply supported beam which is subject to a point load at its mid-span as shown in Figure 1.13(a). Let the value of this load at working condition be P_w. If this load is gradually increased then its value at any other stage can be expressed as λP_w, where λ is a factor known as the load factor. For values of the load above the working load, λ is more than one. As the load factor increases the maximum bending moment in the beam also increases. Eventually at some load factor λ_y, yielding will begin at the outer fibres of the beam. The corresponding maximum elastic bending moments at the working load and at first yield are shown in Figure 1.13(b) and are given respectively as

$$M_w = \frac{P_w L}{4} = \sigma_w Z \qquad (1.20)$$

$$M_y = \frac{\lambda_y P_w L}{4} = \sigma_y Z = \lambda_y \sigma_w Z$$

Further increases of the load factor cause yield to spread sideways along the beam and inwards, towards the neutral axis until

21

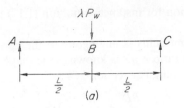

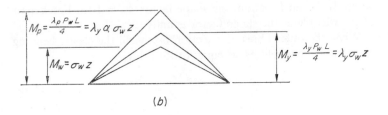

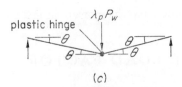

Figure 1.13. Beam subject to point load
(a) Beam and loading
(b) Bending moment diagram
(c) Collapse mechanism

eventually at a load factor λ_p the whole section at mid-span B yields and a plastic hinge is developed there. The beam is thus converted into a mechanism with three hinges. A plastic hinge at the mid-span and two real hinges at the supports. The load factor λ_p cannot be increased any further and the beam collapses with the bending moment at B remaining constant and equal to M_p. This moment is given by

$$M_p = \frac{\lambda_p P_w L}{4} = \sigma_y Z_p \tag{1.21}$$

i.e.

$$\lambda_p P_w = 4M_p/L = 4\sigma_y Z_p/L$$

Furthermore, from equations (1.19) (1.20) and (1.21)

$$\sigma_y Z_p = \lambda_y \sigma_w Z_p = \lambda_y \propto \sigma_w Z$$

22

It follows that

$$M_p = \lambda_y \propto \sigma_w Z \qquad (1.22)$$

Thus

$$\frac{M_p}{M_w} = \lambda_P = \frac{\lambda_y \propto \sigma_w Z}{\sigma_w Z} = \lambda_y \propto \qquad (1.23)$$

This means that the load factor at collapse is given by the product of the load factor at first yield and the shape factor of the section.

1.10. PROPORTIONAL LOADING

Consider now the same simply supported beam but subject to two point loads with working values P_w at mid-span and $3P_w$ at a quarter span as shown in Figure 1.14(a). When gradually increasing these loads, a proportional loading procedure of the beam is achieved if both loads are increased by the same load factor λ and the ratio of the two loads remain constant. The bending moment diagram for this beam during the elastic range is shown in Figure 1.14(b) where it is apparent that yield first takes place at C under the heavier load. Eventually the plastic hinge develops at C and the beam collapses as a mechanism. This is shown in Figure 1.14(c).

The load factor λ_p at collapse is once again obtained by equating the maximum moment in the beam to M_p, thus

$$M_p = \frac{11\lambda_p P_w L}{16}$$

and

$$\lambda_p = \frac{16M_p}{11LP_w}$$

The state of collapse with a plastic hinge at C was obtained by increasing the loads proportionally. If this procedure had not been adopted the collapse stage would have been obtained in a totally different manner. For instance if the loads at C and B are increased by applying load factors λ and λ^1 respectively, with $\lambda^1 = 1 \cdot 5 \lambda$ say, then the bending moments at B and C would both be equal to $3\lambda P_w L/4$. The bending moment over the entire portion BC would also be constant and equal to $3\lambda P_w L/4$. At collapse therefore, this entire zone yields and acts as a plastic hinge as shown in Figure 1.14(d). For values of λ^1 higher than $1\cdot5\lambda$, the bending moment at B becomes greater than that at C and at collapse the plastic hinge develops at B instead of C.

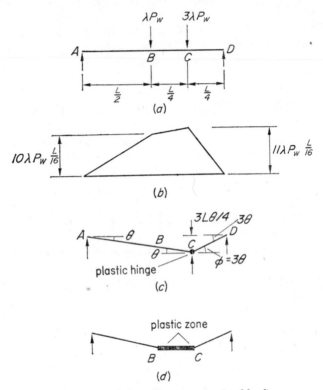

Figure 1.14. Beam subject to proportional loading
(a) Beam and loading
(b) Bending moment diagram
(c) Collapse mechanism
(d) Mechanism with constant moment along BC

This exposé shows clearly that the state of collapse and the position of the plastic hinge is defined entirely by the manner of loading. The plastic theory assumes that the working loads acting on a structure are increased proportionally up to collapse which takes place with the formation of a unique mechanism. In reality structures are not subject to proportional loading. For instance a bridge

24

designed to carry a moving train in gale winds may carry the train with no wind blowing.

In the case of building frames it is more realistic to expect that some of the vertical live loads are already acting, before the frame is subjected to the wind loads. Although it is possible to carry out an elastic-plastic analysis of a frame up to collapse under any loading procedure, it is difficult to predict this loading procedure. Furthermore such an analysis would be more difficult than that for proportional loading. For this reason only the case of proportional loading is discussed in this book. It should be borne in mind however, that while proportional loading is not a necessary procedure in an elastic-plastic analysis, it is a necessary condition for the use of the plastic theory.

1.11. THE USE OF VIRTUAL WORK EQUATION

It has been stated earlier that when the collapse mechanism develops, a structure is statically determinate and the equations of static equilibrium can be utilised to calculate the collapse load factor. It is convenient to derive the equilibrium equation corresponding to a mechanism by the virtual work method.

For instance the mechanism corresponding to the collapse of a simply supported beam loaded at mid-span is shown in Figure 1.13(c). The virtual work equation for this mechanism can be derived by allowing an incremental rotation θ of the collapsing beam. The point of application of the load will move down by an amount $0 \cdot 5 L \theta$ and the work done by this load is $0 \cdot 5 L \theta \cdot \lambda_p P_w$. This work is absorbed by the plastic hinge during the rotation of each half of the beam by θ. Thus the work absorbed by the hinge is $2\theta \cdot M_p$. Equating the work done by the applied load to that absorbed by the hinge rotation, we obtain

$$0 \cdot 5 \lambda_p P_w L \theta = 2 M_p \theta$$

Hence

$$\lambda_p P_w = 4 M_p / L$$

or

$$M_p = \lambda_p P_w L / 4$$

It is seen that the quantity θ appears on both sides of the virtual work equation and thus it does not appear either in the expression for collapse load or for M_p. This is why the value of the incremental rotation is immaterial.

The virtual work equation can be used to derive the load factor that gives rise to any mechanism. In the case of a fixed-ended beam, Figure 1.15, three plastic hinges are required to develop a mechanism. These hinges form at ends A and C of the beam and under the applied load at B. These positions can be easily found

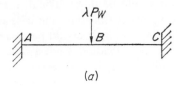

(a)

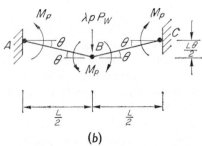

(b)

Figure 1.15. Elastic bending moment diagram for a fixed-ended beam
(a) Beam and loading
(b) Collapse mechanism

from the elastic bending moment diagram, which shows that the bending moments at these points are higher than those anywhere else along the beam.

Once again an incremental rotation θ given to the system will make the applied load do work by $\lambda_p P_w \cdot L\theta/2$. From Figure 1.15(b), it is noticed that the work absorbed by each support hinge is $M_p\theta$, while that of the midspan hinge is $2M_p\theta$. Hence the virtual work equation is

$$M_p\theta + M_p\theta + 2M_p\theta = \lambda_p P_w \cdot L\theta/2$$

i.e.

$$\left. \begin{aligned} \lambda_p P_w &= M_p/L \\ M_p &= \lambda_p P_w L/8 \end{aligned} \right\} \tag{1.24}$$

or

26

Similarly for a propped cantilever supporting a load λP_w at its midspan, the reader can verify that $\lambda_p P_w = 6M_p/L$ and $M_p = \lambda_p P_w/6$. Collapse takes place with two hinges one under the load and the other at the fixed support.

In the case of a fixed ended beam carrying a uniformly distributed load of λW per unit length, W being the value of the working load, collapse also takes place with three plastic hinges A, B and C shown in Figure 1.16. For incremental rotation θ, the total work absorbed

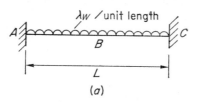

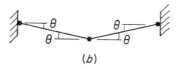

(b)

Figure 1.16. Fixed-ended beam varrying uniformly distributed load
(a) Beam and loading
(b) Collapse mechanism

by these hinges is also $4M_p\theta$. The mid-span point B moves down by $L\theta/2$ and thus the average movement of the total applied load downwards is $L\theta/4$. The work done by the load is thus $\lambda_p WL.L\theta/4$ and the virtual work equation becomes:

$$4M_p\theta = \lambda_p WL^2\theta/4$$
giving $$\lambda_p W = 16M_p/L^2$$
or $$M_p = \lambda_p WL^2/16$$

The collapse mechanism for the simply supported beam in Figure 1.14(a) is shown at Figure 1.14(c). An incremental rotation of portion AC by θ will make point B move down $L\theta/2$ and point C by $3L\theta/4$. Thus the rotation ϕ of portion CD is $3L\theta/4 \div L/4$; i.e. $\phi = 3\theta$. The work done by the point load at B is thus $\lambda_p P_w L\theta/2$

27

and that by the load at C is $3\lambda_p P_w . 3L\theta/4$. The work absorbed by the hinge at C is $M_p(\theta+\phi)$ and the virtual work equation is:

$$4M_p\theta = \lambda_p P_w L\theta/2 + 3\lambda_p P_w . 3L\theta/4$$

$$\therefore \lambda_p P_w = 16M_p/11L$$

or $$M_p = 11\lambda_p P_w L/16$$

In the above instance it was easy to find the position of the plastic hinges and hence the collapse mechanism. Sometimes the exact positions of all the hinges are not apparent. For instance, consider the propped cantilever shown in Figure 1.17 which is carrying a

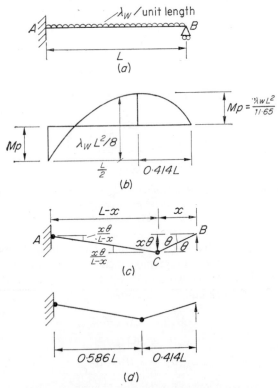

Figure 1.17. Beam with a simply supported end
(a) Beam and loading
(b) Elastic bending moment diagram
(c) A possible collapse mechanism
(d) The actual collapse mechanism

28

uniformly distributed load λW per unit length. This beam becomes a mechanism with two hinges. One hinge develops at end A where the hogging bending moment is maximum. The other hinge develops in the span at some point C where the sagging moment is maximum. To find the position of this hinge, let us assume that it is at a distance x from the simply-supported end as shown in Figure 1.17(c). Giving the portion BC an incremental rotation θ, point C moves down by $x\theta$. Hence the average movement of the uniform load down is $x\theta/2$ and the hinge rotation at A is $x\theta/(L-x)$. At collapse the work done by the load is thus $\lambda_p WL.x\theta/2$ and the work absorbed by the hinges at A and C are $x\theta M_p/(L-x)$ and $M_p[\theta + x\theta/(L-x)]$ respectively. The virtual work equation is thus given by

$$\frac{\lambda_p WLx\theta}{2} = M_p\theta + \frac{2M_p x\theta}{L-x}$$

i.e.
$$\lambda_p = \frac{2M_p}{LW}\left(\frac{L-x}{Lx-x^2}\right)$$

The beam collapses when λ_p is a minimum. That is to say as the load factor increases collapse takes place with a mechanism that can develop with the smallest load. The second hinge can be anywhere in the span so long the load that causes it is the least. This is not only common sense, but in fact it is also in accordance with the 'Kinematic Theorem' which was stated earlier. Thus at collapse

$$\frac{d\lambda_p}{dx} = \frac{2M_p}{WL}\left[\frac{xL-x^2-(L+x)(L-2x)}{(Lx-x^2)^2}\right] = 0$$

which gives $x = 0.414L$. Hence the collapse load factor is given by

$$\lambda_p W = 11.65M_p/L^2$$

i.e.
$$M_p = \lambda_p WL^2/11.65. \tag{1.25}$$

The actual collapse mechanism is shown in Figure 1.17(d). The procedure just described for finding the position of a plastic hinge in a beam which is uniformly loaded can be applied to other structures such as continuous beams and frames.

1.12 COLLAPSE OF CONTINUOUS BEAMS

In order to calculate the collapse load of a continuous beam with several spans, the virtual work equation can be used to calculate the collapse load of each span separately. The collapse load of the beam is then selected to be the lowest load that causes the collapse of the weakest span in the beam.

Consider, for instance, the uniform continuous beam $ABCDEF$ of Figure 1.18. The spans AB and BD are of length L, while span DF is $2L$. AB carries a total uniform working load, W, BD carries a working point load W acting at its midspan and DF carries a working load of $2W$ acting at a distance of $L/2$ from support D. As the loads are increased proportionally by multiplying each span load by the same load factor λ, the first collapse mechanism may develop at any of the three spans. These mechanisms are shown in Figures 1.18(c), (d) and (e). For the mechanism to develop in span AB, Figure 1.18(c), a hinge has to form at B, where the hogging bending moment is highest in this span, and another hinge has to form somewhere such as G in the span where the sagging moment is the highest. This type of collapse is similar to the propped cantilever of Figure 1.17, analysed in the last section and hence point G is $0\cdot414L$ from the simple support A. Once again the load factor is given by equation (1.25), thus

$$\left.\begin{array}{l} \lambda_p = 11\cdot65M_p/LW \\ M_p = \lambda_pWL/11\cdot65 \end{array}\right\} \tag{1.26}$$

For the mechanism to develop in span BD, the hogging bending moments at B and C should be numerically equal to the sagging bending moment at mid-span C and they all take the numerical value M_p. This is similar to the case of the fixed-ended beam of Figure 1.15 analysed in the last section and hence the collapse load factor is given by equation (1.24),

i.e.
$$\left.\begin{array}{l} \lambda_p = 8M_p/WL \\ M_p = \lambda_pWL/8 \end{array}\right\} \tag{1.27}$$

Finally for the mechanism to develop in span DF, the numerical values of M_D and M_E should be equal to M_p. This mechanism is shown in Figure 1.18(e). The collapse load factor λ_p with this mechanism can be derived by using the virtual work equation. Giving an incremental rotation θ to the portion DE, then point E moves down by $\theta \cdot L/2$ and the work done by the load $2W\lambda_p$ is $\lambda_p WL\theta$. The rotation ϕ of the simply supported end F is $0 \cdot 5L\theta/1 \cdot 5L$, i.e. $\phi = \theta/3$. The hinge rotation at E is therefore $\theta + \theta/3$ and the work

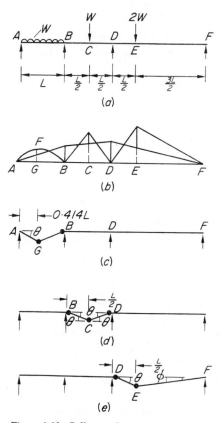

Figure 1.18. Collapse of a continuous beam
(a) Beam and loading
(b) Elastic bending moment diagram
(c) Mechanism in AB alone
(d) Mechanism in BD
(e) Mechanism in DF

absorbed by this hinge is $(\theta + \theta/3)M_p$. The work absorbed by the hinge at D is $M_p\theta$ and the virtual work equation becomes:

$$M_p\theta + M_p(\theta + \theta/3) = \lambda_p WL\theta$$

31

Hence

$$\left.\begin{array}{l} \lambda_p = 7M_p/3WL \\ M_p = 3\lambda_pWL/7 \end{array}\right\} \qquad (1.28)$$

The continuous beam collapses when only one of these mechanisms develops. The lowest of the load factors given by equations (1.26), (1.27) and (1.28) is the load factor that causes collapse. Since

$$11{\cdot}65M_p/WL > 8M_p/WL > 7M_p/3WL,$$

it follows that collapse takes place by the mechanism of Figure 1.18(e) with $\lambda_p = 7M_p/3WL$ given by equation (1.28).

1.13. SELECTION OF MECHANISMS

In the case of continuous beams, although there is the possibility of the development of a separate mechanism in each span, the final collapse takes place with only one of these mechanisms. In the case of frames, however, there may be the possibility of several separate member mechanisms. Collapse may take place by any one of these individual mechanisms. On the other hand hinges may be selected throughout the frame that render the frame or part of it into a mechanism. In a frame collapsing in this manner, no member will have its own individual mechanism. In order to find the collapse load factor λ_p all these mechanisms have to be investigated to find out the one that can develop with the lowest load factor. To achieve this aim, it is necessary to list all the possible positions of hinges in a frame and select from this list two or more hinges that can give rise to the development of a mechanism.

For a member that is not loaded anywhere along its span, there is the possibility of a hinge forming at either end of such a member. If, however, the member is subject to some form of lateral loading, then an extra hinge can form along the span of the member. A mechanism in the frame is uniquely defined when all the incremental rotations in the frame can be algebraically expressed in terms of one of these rotations.

As an example consider the portal frame of Figure 1.19. The columns AB and ED are not loaded along their lengths, hinges can therefore develop only at their ends. On the other hand the beam BCD is laterally loaded at C and can therefore have three hinges. Altogether seven hinges can develop in this frame. In Figure 1.19(b)

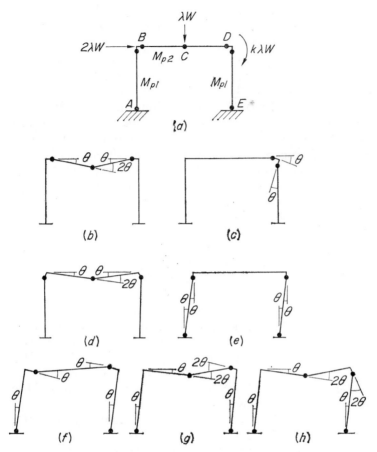

Figure 1.19. Separate mechanisms in a portal frame

(a) Possible position of hinges
(b) Beam mechanism
(c) Joint mechanism
(d) Beam mechanism

(e) Sway mechanism
(f) Sway mechanism
(g) Combined mechanism
(h) Combined mechanism

a beam-type collapse mechanism is shown with hinges B, C and D in the beam and an incremental rotation θ. The mechanism shown in Figure 1.19(d) is also beam type except the hinges at B and D are developed at the columns instead of the ends of the beams. Whether the hinges form at the end of the beam or the column heads depends on the values of the fully plastic hinge moments M_{p2} of the beam and M_{p1} of the columns. If $M_{p2} > M_{p1}$ then the hinges develop in the columns, otherwise they form at the ends of the beam. In Figure 1.19(c) a joint mechanism is developed. Such a mechanism is possible at a joint where an external moment is acting and it takes place with a plastic hinge at the end of all the members meeting at that joint. Mechanisms e and f in the figure are sway type mechanisms. These can generally develop when the side load acting on the frame is large. The last two mechanisms are a combined type mechanism with a hinge developing under the point load at C and the frame moving sideways. In fact, a combined mechanism can be obtained by superimposing a beam type mechanism on a sway mechanism. For instance the hinge rotations at end B of the beam in Fugures b and f are in opposite directions. Once these two mechanisms are added together, the hinge rotation at B reduces to zero. The hinge rotations at D are in the same direction in both these mechanisms and therefore they add up, resulting in a rotation of 2θ at D. In this manner mechanism g is obtained.

For each one of these mechanisms a virtual work equation can be written and the load factor to cause collapse by each mechanism can be calculated. The lowest of these factors is the actual collapse load factor. It is not necessary to try every mechanism, since if M_{p2} is different from M_{p1} then the hinges at B and D may form either in the beam or the column. Of course the mechanisms satisfying the actual conditions are the ones that are possible to develop. If $M_{p1} = M_{p2}$, then the load factors given by mechanisms b and d are equal. Similarly the load factors for mechanisms e and f will be equal and so will those for mechanisms g and h.

Let us assume, by way of an example, that $M_{p2} = 1 \cdot 5 M_{p1}$, $AB = L$, $BC = 3L/4 = CD$ and $k = 0$. Since $M_{p1} < M_{p2}$ mechanisms d, e, and h are the only ones that are possible to develop. In the case of mechanism d, the reader can derive the virtual work equation as

$$2M_{p1}\theta + 3M_{p1}\theta = 3\lambda_p WL\theta/4, \qquad (M_{p2} = 1 \cdot 5 M_{p1});$$

and
$$\left. \begin{array}{l} \therefore \ \lambda_p = 20M_{p1}/3WL \\ M_{p1} = 3\lambda_p WL/20 \end{array} \right\} \qquad (1.29)$$

The virtual work equation for mechanism e is

$$4M_{p1}\theta = 2\lambda_p WL\theta$$

and
$$\left. \begin{array}{l} \therefore \ \lambda_p = 2M_{p1}/WL \\ M_{p1} = \lambda_p WL/2 \end{array} \right\} \qquad (1.30)$$

34

Finally, for mechanism h the virtual work equation is

$$2\theta M_{p1} + 2\theta M_{p1} + 2\theta M_{p2} = 2\lambda_p WL\theta + 3\lambda_p WL\theta/4$$

This gives

i.e.
$$\left. \begin{array}{l} \lambda_p = 28M_{p1}/11WL \\ M_{p1} = 11\lambda_p WL/28 \end{array} \right\} \tag{1.31}$$

From equations (1.29) through (1.31) the smallest value of λ_p is $2M_{p1}/WL$ given by mechanism e which is therefore the collapse mechanism.

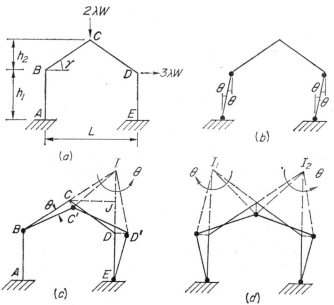

Figure 1.20. Pitched roof frame

(a) Frame and loading   (c) Frame mechanism
(b) Sway mechanism   (d) Symmetrical frame mechanism

As another example consider the pitched roof frame of Figure 1.20. The frame is fixed at A and E and is subject to a vertical load of $2\lambda W$ at the apex C and a horizontal load of $3\lambda W$ at D. The loads $2W$ and $3W$ are the working loads. The members of the frame are made out of the same cross-section with the fully plastic hinge moment M_p. This frame may collapse with one of the three mechanisms shown in the figure. In the case of the sway mechanism, Figure 1.20(b) hinges form at A, B, D and E. The hinges at B and D

may form either in the columns or in the rafter. For an incremental rotation θ, as shown in the figure, the work absorbed by the hinges is $4M_p\theta$. The vertical load at C does not move down and hence does no work. The horizontal load at D moves by $h_1\theta$ and hence the work done by it is $3\lambda Wh_1\theta$.

The virtual work equation is thus

$$4M_p\theta = 3\lambda_p Wh_1\theta$$

i.e.

$$\left.\begin{array}{l} \lambda_p = \dfrac{4M_p}{3Wh_1} \\[2ex] M_p = 3\lambda_p Wh_1/4 \end{array}\right\} \tag{1.32}$$

or

In the case of the frame mechanism, Figure 1.20(c) hinges develop at B, C, D and E. In order to find the hinge rotations, member CD is rotated by θ about the instantaneous centre I. From the geometry of the frame, it is evident that $CJ = L/2$, $JD = h_2 = = IJ$ and $CI = CB$. Thus $CC' = IC\theta$ and the vertical movement of the load $2\lambda W$ is $CC' \cos \gamma = L\theta$. The hinge rotation at B is $CC'/BC = \theta$. The horizontal movement at D is $2h_2\theta$ and the hinge rotation at E is therefore $2h_2\theta/h_1$. Since members CB and CD both rotate by θ, the hinge rotation at C is, therefore, 2θ. On the other hand the hinge at D rotates by $\theta+2h_2\theta/h_1$. The virtual work equation is thus given by

$$M_p\left[\theta+2\frac{h_2}{h_1}\theta+2\theta+\left(\theta+\frac{2h_2\theta}{h_1}\right)\right] = 2\lambda_p WL\theta+3\lambda_p W \cdot 2h_2\theta$$

Hence for $k = h_2/h_1$ we obtain

$$\left.\begin{array}{l} \lambda_p = M_p(1+k)/2WL \\[1ex] M_p = 2\lambda_p WL/(1+k) \end{array}\right\} \tag{1.33}$$

and

The reader can derive the virtual work equation for the symmetrical frame mechanism of Figure 1.20(d). This gives

$$\left.\begin{array}{l} \lambda_p = M_p(1+2k)/2WL \\[1ex] M_p = 2\lambda_p WL/(1+2k) \end{array}\right\} \tag{1.34}$$

and

It is evident from equations (1.33) and (1.34) that, because $(1+2k) > (1+k)$, the symmetrical frame mechanism does not develop in this frame under the defined loads. On the other hand comparing equations (1.32) and (1.33), it is noticed that the frame collapses with the sway mechanism of Figure 1.20(b), if $8L < 3(h_1+h_2)$. Otherwise the frame mechanism of Figure 1.20(c) gives the collapse load.

36

1.14. DESIGN BY RIGID-PLASTIC THEORY

It has been pointed out that when a rigid-plastic collapse of a structure takes place, the structure is statically determinate. This simplifies the design procedure and reduces it to that of a simple analysis. In general, the problem of design is the reverse of the problem of an analysis. In the case of an analysis, the loads, dimensions and sectional properties are known and it is required to find the member forces and bending moments.

In the problem of design, the sectional properties are required so that the structure can carry the applied loads. For statically determinate structures an analysis can be carried out without a knowledge of sectional properties. Once this is done, suitable sections can be selected to withstand the member forces and moments. For this reason in the preceding sections, the result of an analysis was always expressed either to predict the load factor or to calculate the fully plastic moments of the section.

Various codes of practice decide upon the load factor λ_p at collapse in a definite manner. For intance, at present, the load factor at collapse under vertical loadings alone is 1·75, while that und er combined vertical and wind loads is 1.4. Once the type of loading and the corresponding load factor is decided upon, a rigid plastic analysis of a structure determines the sectional properties of the members. In the case of the simply supported beam of Figure 1.13, for instance, the virtual work equation gave the value of M_p as $\lambda_p P_w L/4$. For a load factor $\lambda_p = 1·75$, once the applied working loads P_w an d the span L of the beam are decided upon, the required M_p for the section can be calculated. This value of M_p together with the yield stress σ_y of the material are used to calculate Z_p of the section from

$$Z_p = M_p/\sigma_y$$

A section is then selected with an actual Z_p greater than that calculated. This same procedure can be used to design propped cantilevers and fixed-ended beams.

In the case of continuous beams it is possible to obtain more than one acceptable design depending on the type of mechanisms that are allowed to develop at collapse and the type of sections to be used. For instance consider the design of the continuous beam of Section

1.12 and shown in Figure 1.18. There are three possible types of mechanisms that can develop when the beam collapses. The value of the fully plastic hinge moment for these mechanisms are given by equations (1.26), (1.27), and (1.28).

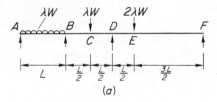

(a)

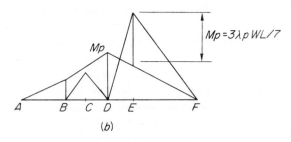

(b)

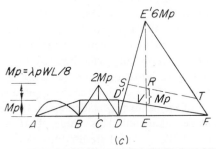

(c)

Figure 1.21. Bending moment diagram for continuous beam
(a) Beam and loading
(b) Bending moment diagram when mechanism develops in DF
(c) Bending moment diagram when mechanism develops in BD

If it is required to make the beam out of the same uniform section then it is necessary to make this section sufficiently strong to withstand the largest M_p given by these three equations. Since $3\lambda_p WL/7 > \lambda_p WL/8 > \lambda_p WL/11\cdot65$, it follows that the selected section should have $M_p > 3\lambda_p WL/7$. The bending moment diagram for the beam with $M_p = 3\lambda_p WL/7$ is shown in Figure 1.21(b), and the collapse mechanism is shown in Figure 1.18(e). The free bending moment diagram at C is $7M_p/24$ and that halfway between A and B is $7M_p/24$. At no point along ABD, does the bending moment reach the value of M_p. Thus so long as the section selected is sufficiently large to withstand a moment $3\lambda_p WL/7$ at D and E, it will be safe to carry the loads on all the spans.

A lighter design of this beam can be obtained by forcing the beam to collapse with a mechanism in span BD and another one in span DF. From equation (1.27), $M_p = \lambda_p WL/8$ obtained by allowing a mechanism to develop in span BD, Figure 1.18(d). Three hinges would develop at B, C and D. The hogging moments at the supports B and D will be M_p, the free moment at C will be $2M_p$ thus the sagging moment at C will be M_p. In Figure 1.21(c), the bending moment diagram for the beam is shown with these moments indicated. Now if

$$M_p = \lambda_p WL/8$$

then

$$\lambda_p W = 8M_p/L$$

and with load acting on span AB, the free bending moment midway between A and B is

$$M = \lambda_p WL/8 = \frac{L}{8} \cdot \frac{8M_p}{L} = M_p$$

Thus if the same section was used for the beam to withstand $M_p = \lambda_p WL/8$, then such a section would not collapse by the mechanism shown in Figure 1.18(c), developing in span AB. Everywhere in this span the bending moment is less than M_p as shown in Figure 1.21(c).

For span DF, however, the free bending moment at E is

$$M_{\text{EFREE}} = 2\lambda_p W \cdot 0{\cdot}5L \cdot 1{\cdot}5L/2L = \frac{3}{4} L \cdot \frac{8M_p}{L} = 6M_p$$

Since the support moment $M_D = M_p$ it follows that the net sagging moment at E is

$$M_E = 6M_p - 1{\cdot}5M_p/2 = 21M_p/4$$

This indicates that the section selected to withstand the moments in the middle span is not sufficiently strong to carry the sagging moment at E. In Figure 1.21(c) the straight line ST is drawn parallel to $D'F$ so that $RV = M_p$. The uniform section used for AD if continued to F will provide a plastic moment of resistance of $M_p = RV$

39

all along *DF*. The portion of the beam vertically below *ST*, will thus require reinforcement. This can be provided by a pair of parallel plates covering the uniform section between *S* and *T* and having a moment of resistance equal to *RE'*, i.e. $17M_p/4$. A beam designed in this manner will collapse with mechanisms developing in spans *BD* and *DF* simultaneously.

1.15. EFFECT OF AXIAL LOAD ON M_p

In general the members of a structure are subject to axial forces as well as bending moments. The effect of the axial load in a member is to reduce its fully plastic moment. Consider a member subject to a moment and a force *P* with a yield stress σ_y. When the section of this member becomes fully plastic, this takes place by the combined effect of the moment and the axial force. Assuming that the axial force is acting centrally, the area of the section which yields because of the presence of *P* is

$$A_p = P/\sigma_y \qquad (1.35)$$

The rest of the section yields by the acting moment which causes tension and compression on equal areas on either side of the plastic neutral axis. These stresses are represented diagrammatically in Figure 1.22. The central section of area A_p reaches the stress σ_y by the axial load as shown in Figure 1.22(b). The section loses the fully plastic hinge moment M_{pp} of this central section and the bending stress distribution on the rest of the section is shown in Figure 1.22 (c). The final fully plastic stress distribution is the sum of these two.

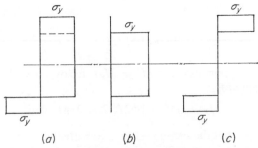

Figure 1.22. Effect of axial loads
(a) Final stresses
(b) Stresses caused by P
(c) Stresses caused by the moment

This is shown in Figure 1.22(a). Naturally the reduced plastic hinge moment M_p' of the section is given by

$$M_p' = M_p - M_{pp} \qquad (1.36)$$

For example, let a rectangular section of width B and depth D be subject to an axial force P. The area A_p is given by $P/\sigma_y = Bd$, where d is the depth of the central portion yielded by the axial force, i.e.

$$d = P/B\sigma_y$$

The plastic section modulus of the central portion Z_{pp} is therefore given by

$$Z_{pp} = Bd^2/4 = P^2/(4B\sigma_y^2)$$

The reduced plastic section modulus is thus given by

$$Z_p' = Z_p - Z_{pp} = \frac{BD^2}{4} - \frac{P^2}{4B\sigma_y^2}$$

and the reduced plastic hinge moment $M_p' = \sigma_y Z_p'$ is given by

$$M_p' = \sigma_y \left[\frac{BD^2}{4} - \frac{P^2}{4B\sigma_y^2} \right]$$

$$\therefore \ M_p' = M_p(1 - d^2/D^2)$$

In a similar manner it can be found that the reduced plastic hinge moment of the I section of Section 1.7, Figure 1.12, is given by

$$M_p' = M_p \{1 - 0 \cdot 25 t_1 d^2 / [Bt_2(D - t_2) + 0 \cdot 25(D - 2t_2)^2 t_1]\},$$

where d is the depth of the central portion of the web which is yielded by the effect of an axial load P. The value of d is again calculated using equation (1.35). In this expression for M_p' it is assumed that the central area, which is yielded by the axial load, does not spread to the flanges of the I section. Generally the reduction in fully plastic moment due to an axial force is more severe for an I section than for a rectangular section.

The application of the plastic theory is limited to beams and small structures such as building frames with one or two storeys. As a rule the axial forces in the members of these structures are small and therefore, their effects on reducing the fully plastic hinge moments of the members are also small. In tall building frames, with stanchions carrying heavy axial thrusts, the effect of thrusts on reducing the values of M_p in these stanchions can be considerable. This is particularly the case with the stanchions of the lower storeys. Furthermore such structures require elastic-plastic failure load analyses, where instability effects due to high axial loads are also important.

In such an analysis, therefore, the effect of axial forces on reducing the fully plastic moment of the section becomes very significant. This is particularly so after the formation of a plastic hinge in a member. As the load factor increases, the axial forces in the members also increase and the combined effect of instability and reduction in M_p become major factors in producing failure in the structure. These factors are fully taken into consideration in Chapters 3 to 8.

EXERCISES

1. Using the slope-deflection method, calculate the bending moments M_{AB}, M_{BA}, M_{BC} and M_{CB} for the uniform fixed ended column ABC in Figure 1.23 when an external moment M is applied at B.

Answer: $0.5M/9$, $4M/9$, $3M/9$

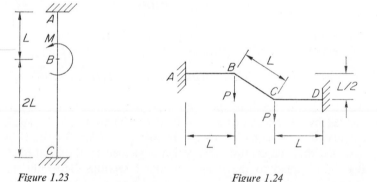

Figure 1.23   Figure 1.24

2. Using the slope deflection equations, calculate the downward vertical movements and joint rotations at B and C of the rigidly jointed structure in Figure 1.24. EI is constant.

Answer: $v = PL^2/6EI$, $\theta_B = PL/6EI$, $\theta_c = -PL/6EI$

3. A uniform simply supported beam of length L, full plastic moment M_p and unit shape factor is subject to a variable point load at its mid-span. At a certain value of the load, the maximum mid-span bending moment becomes equal to M_p and the beam begins to behave as a mechanism. Show that the central deflection of the beam at this instant is given by $M_pL^2/12EI$.

4. A uniform beam of span L, full plastic moment M_p and unit shape factor is built in at both ends and carries a uniformly distributed load. Increasing the load causes the formation of two plastic hinges at the ends of the beam when the bending moments there reach the value M_p. Show that the central deflection of the beam at this instant is given by $M_pL^2/32EI$.

5. Further increases of the load in the last example would finally cause the formation of a third and final hinge at mid-span. What would be the value of the central deflection?

Answer: $M_pL^2/12EI$.

6. The two bay rectangular portal of Figure 1.25 has three equal columns of fully plastic hinge moment M_p and a continuous beam of fully plastic hinge moment $2M_p$. Calculate the plastic collapse load factor.

Answer: $\lambda_p = 11M_p/4WL$

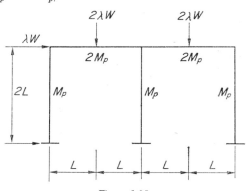

Figure 1.25

7. Calculate the rigid-plastic collapse load factor λ_p of the pinned base pitched roof frame shown in Figure 1.26. All the members are made out of the same section with constant M_p.

Answer: $\lambda_p = 5M_p/2WL$

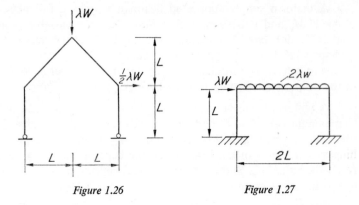

Figure 1.26   *Figure 1.27*

8. The fully plastic hinge moment for the members of the portal shown in Figure 1.27 is M_p. The frame is subject to a side load of λW and a total uniform load of $2\lambda W$. Calculate the rigid plastic collapse load factor.

Answer: $\lambda_p = 2{\cdot}96 M_p/WL.$

44

CHAPTER 2

Matrix Methods of Structural Analysis

2.1. INTRODUCTION

Matrix methods invariably appear to be rather cumbersome particularly in dealing with simple structural problems. In reality, however, they are straightforward and form a powerful tool for dealing with large and complex structures. They become especially useful when a computer is used to perform the numerical operations. Although any technique of structural analysis such as the moment distribution method, for instance, can be performed using matrix algebra, there are basically two main matrix methods that have been well established in the field of structural analysis.

The first method, which is used more widely than the second, is the matrix displacement method which expresses the internal member forces in terms of the joint displacements and then proceeds to solve a set of joint equilibrium equations in order to determine the unknown displacements. This method is popular because it does not involve the concept of redundancies and deals with statically determinate and indeterminate structures with equal efficiency.

The second method is the matrix force method. This expresses the conditions of equilibrium between the externally applied loads and the resulting internal forces and solves the compatibility equations for the unknown redundants. Recently, and particularly after the development of automatic devices for the construction of equilibrium equations, the matrix force method has drawn more attention. This method involves the solution of a smaller number of equations, one per unknown redundant, and can be used as a powerful tool in the design of structures where deflections are part of the design criteria.

2.2. MATRIX DISPLACEMENT METHOD

2.2.1. The assemblage of member stiffness matrices

(i) For pin-jointed structures

The member *AB* of a pin-jointed structure, shown in Figure 2.1, is of area *A*, length *L* and modulus of elasticity *E*. It is subject to an axial force *p* that makes the member extend by an amount *u*. It

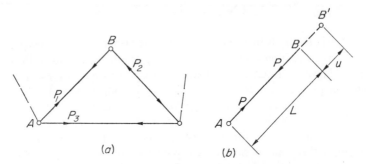

Figure 2.1. *Effect of force on a pin-jointed structure*
(a) Part of a pin-jointed structure
(b) Extension of a typical member

follows directly from Hooke's law that this force can be expressed in terms of the extension of the member as

$$p = \frac{EAu}{L} \qquad (2.1)$$

where the term EA/L is the axial stiffness of that member. Similarly in the pin-jointed structure the forces and the extensions (or contractions) can be expressed as

$$p_1 = E_1 A_1 u_1 / L_1$$
$$p_2 = E_2 A_2 u_2 / L_2 \qquad (2.2)$$
$$p_i = E_i A_i u_i / L_i$$
ect.

writing these equations in matrix form

$$
\begin{bmatrix} p_1 \\ p_2 \\ \cdot \\ \cdot \\ p_i \\ \cdot \\ p_n \end{bmatrix}
=
\begin{bmatrix}
E_1 A_1 / L_1 & 0 & & & \\
0 & E_2 A_2 / L_2 & & & \\
\cdot & \cdot & \cdot & & \\
\cdot & \cdot & \cdot & & \\
0 & 0 & E_i A_i / L_i & & \\
\cdot & \cdot & \cdot & \cdot & \\
0 & 0 & 0 & E_n A_n / L_n
\end{bmatrix}
\begin{bmatrix} u_1 \\ u_2 \\ \cdot \\ \cdot \\ u_i \\ \cdot \\ u_n \end{bmatrix}
\qquad (2.3)
$$

or simply $P = k \cdot Z$ \qquad (2.4)

In equation (2.4) the load vector $P = \{p_1 p_2 \ldots p_n\}$ summarises all the member forces in the structure, i being a typical member and n being the last. The vector $Z = \{u_1 \ldots u_n\}$ summarises the accompanying displacements of the members and the matrix k is the member stiffness matrix of the structure. The term 'displacement' is used to include all member distortions such as extension, shortening or rotation.

(ii) For a rigidly jointed plane frame

In Figure 2.2 a member of a rigidly jointed plane frame is shown subject to 'member forces', that is to say loads and/or moments. These forces cause an extension u, a sway v of one end relative to the other and end rotations θ_{AB} and θ_{BA} at the ends A and B respectively. The sign conventions for forces and displacements are also shown in the figure. The slope-deflection equations for this member are

$$M_{AB} = \frac{-6EI}{L^2} v + \frac{4EI}{L} \theta_{AB} + \frac{2EI}{L} \theta_{BA}$$

$$\qquad (2.5)$$

$$M_{BA} = \frac{-6EI}{L^2} v + \frac{2EI}{L} \theta_{AB} + \frac{4EI}{L} \theta_{BA}$$

47

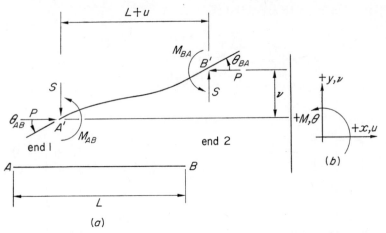

Figure 2.2. *Deformation of a member of a rigidly-jointed plane frame*
(a) End forces and displacements
(b) Sign convention

Taking moments about point B', ignoring pv term, the shear force S is obtained

$$S = \frac{12EI}{L^3} v - \frac{6EI}{L^2} \theta_{AB} - \frac{6EI}{L^2} \theta_{BA} \qquad (2.6)$$

Writing equation (2.1) for the axial stiffness of the member together with equations (2.6) and (2.5) in matrix form we obtain

$$
\begin{bmatrix} p_{AB} \\ S_{AB} \\ M_{AB} \\ M_{BA} \end{bmatrix}
=
\begin{bmatrix}
EA/L & 0 & 0 & 0 \\
0 & 12EI/L^3 & -6EI/L^2 & -6EI/L^2 \\
0 & -6EI/L^2 & 4EI/L & 2EI/L \\
0 & -6EI/L^2 & 2EI/L & 4EI/L
\end{bmatrix}
\begin{bmatrix} u_{AB} \\ v_{AB} \\ \theta_{AB} \\ \theta_{BA} \end{bmatrix}
$$

$$(2.7)$$

or just $\qquad\qquad P_{AB} = k_{AB} . Z_{AB} \qquad\qquad (2.8)$

where the vector P_{AB} summarises the forces in member AB, Z_{AB} summarises the displacements and the matrix k_{AB} is the stiffness

48

matrix for the member. Denoting for simplicity,

$$a = EA/L$$
$$b = 12EI/L^3$$
$$d = -6EI/L^2 \qquad (2.9)$$
$$e = 4EI/L$$
$$f = 0 \cdot 5e,$$

the stiffness matrix of the member becomes

$$k_{AB} = \begin{bmatrix} a & & \text{symmetrical} \\ 0 & b & & \\ 0 & d & e & \\ 0 & d & f & e \end{bmatrix} \qquad (2.10)$$

Equations similar to (2.7), (2.8) or (2.10) can be obtained for all the other members of the plane frame. When these are compounded together, the member stiffness matrix of the frame is formed. This consists of submatrices similar to (2.10) whose leading diagonals form the leading diagonal of the assembled matrix with zeros elsewhere thus

$$\begin{bmatrix} P_1 \\ P_2 \\ P_i \\ P_n \end{bmatrix} = \begin{bmatrix} k_1 & & & \\ & k_2 & & \\ & & k_i & \\ & & & k_n \end{bmatrix} \begin{bmatrix} Z_1 \\ Z_2 \\ Z_i \\ Z_n \end{bmatrix} \qquad (2.11)$$

or

$$P = k \cdot Z \qquad (2.12)$$

where $P_i = \{p_{AB} \ S_{AB} \ M_{AB} \ M_{BA}\}_i$ are the forces in a typical member i and Z_i are the corresponding displacements. The frame has n members and equation (2.12) expresses the stiffness of all these members.

(iii) For a rigidly jointed space frame

A member of a rigidly jointed space frame, Figure 2.3, that has an axis of symmetry bends about two perpendicular axes. Equations, similar to (2.5) and (2.6) can be written for bending

5

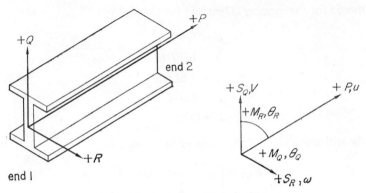

Figure 2.3. A member of a space frame
(a) Member axes
(b) Sign convention for member forces and displacements

about both these axes. The member also twists about its longitudinal axis and the torsional stiffness of the member is given by

$$T = \frac{GJ}{L}\alpha \qquad (2.13)$$

where T is the applied torque, α is the angle of twist and GJ is the torsional rigidity of the cross-section. When writing these equations together they become

$$
\begin{bmatrix} p \\ S_Q \\ M_{R1} \\ M_{R2} \\ T \\ S_R \\ M_{Q1} \\ M_{Q2} \end{bmatrix}
=
\begin{bmatrix}
a & & & & & & & \\
0 & b_R & & & \text{symmetrical} & & & \\
0 & d_R & e_R & & & & & \\
0 & d_R & f_R & e_R & & & & \\
0 & 0 & 0 & 0 & g & & & \\
0 & 0 & 0 & 0 & 0 & b_Q & & \\
0 & 0 & 0 & 0 & 0 & d_Q & e_Q & \\
0 & 0 & 0 & 0 & 0 & d_Q & f_Q & e_Q
\end{bmatrix}
\begin{bmatrix} u \\ v \\ \theta_{R1} \\ \theta_{R2} \\ \alpha \\ w \\ \theta_{Q2} \\ \theta_{Q1} \end{bmatrix}
\qquad (2.14)
$$

where $g = GJ/L$, a, b, d and e are the quantities defined by equations (2.9), S_Q and M_Q are shear along and bending moment about the Q axis of the member. The suffixes Q and R are for the Q and R axes of the member and the second suffix associated with the moments and rotations is to identify end 1 and 2 of the member. The sign convention is shown in Figure 2.3(b).

50

2.2.2. Transformation of displacements

So far, the equations presented in the last section give the member forces in a structure in terms of the displacements in every separate member. In general, however, it is more convenient to express these in terms of the joint displacements and along a set of reference axes suitably chosen for the structure as a whole. The displacement transformation matrix for a member expresses the displacements at the ends of the member, in its local co-ordinates.

In order to construct this matrix for a member of a pin-jointed plane frame let X and Y be the overall reference axes and P and Q be the axes of the member. The member itself is specified by points A and B. Point A being at the first end of the member and point B at its second end so that the positive direction of the P axis is from A to B. Point C is on the Q axis of the member and the positive direction of this axis is from A to C. The sign convension is shown in Figure 2.4(b). Let the co-ordinates of points A and B be (X_1, Y_1) and (X_2, Y_2) respectively. These are shown in Figure 2.4 where the positive direction of the P axis of the member is indicated by an arrow the head of which is pointed to the second end of the member. The length L of the member is given by

$$L = \sqrt{[(X_2-X_1)^2+(Y_2-Y_1)^2]} \qquad (2.15)$$

The angles α and β are between the positive P axis of the member and the positive X and Y axes respectively. The direction cosines l_p and m_p of the members are given by

$$l_p = \cos \alpha = (X_2-X_1)/L, \qquad (2.16)$$
$$m_p = \cos \beta = (Y_2-Y_1)/L, \qquad (2.17)$$

Let the first end of the member be connected to joint i in the frame and its second end to joint j. Due to some externally applied forces, let the displacements of these joints in the overall reference co-ordinates be

$$X_i = \{x \ y\}_i, \qquad (2.18)$$
$$X_j = \{x \ y\}_j, \qquad (2.19)$$

As a result of these joint displacements the ends of the member move by u_1 and u_2 along its P axis, so that the new position of the member is given by A^1 and B^1 as shown in Figure 2.4(c). The

5*

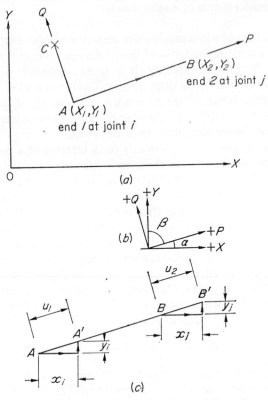

Figure 2.4. Displacements of a pin-ended member
(a) Member co-ordinates
(b) The sign convention
(c) Member and joint displacements

extension of the member will be

$$u = u_2 - u_1 \qquad (2.20)$$

but

$$u_1 = l_p x_i + m_p y_i \qquad (2.21)$$

and

$$u_2 = l_p x_j + m_p y_j \qquad (2.22)$$

it follows that:

$$u = [\underset{\text{at joint } i}{-l_p \; -m_p} \ldots \underset{\text{at joint } j}{l_p \; m_p}] \{x_i \; y_i \ldots x_j \; y_j\} \qquad (2.23)$$

The row vector $[-l_p \; -m_p \ldots l_p \; m_p]$ is the displacement transformation matrix of the member.

Similarly the extension u of a member in a pin-jointed space frame is given by

$$u = [-l_p \ -m_p \ -n_p \ldots \overset{\text{at joint } i}{} \ l_p \ m_p \ n_p] \{x_i \ y_i \ z_i \ldots x_j \ y_j \ z_j\} \quad (2.24)$$

where z_i and z_j are the displacements of joints i and j respectively parallel to Z axis of the structure, n_p is the direction cosine of the third axis R of the member with respect to the Z axis and given by

$$n_p = (Z_2-Z_1)/L \quad (2.25)$$

where Z_2 and Z_1 are the Z co-ordinates of the first and second ends of the space member.

In the case of a member of a rigidly-jointed plane frame as shown in Figure 2.5 in the XY plane, the extension u of this member is

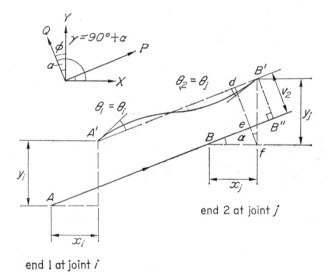

end 1 at joint i

Figure 2.5. Displacements of a member in a rigidly-jointed plane frame

also given by equation (2.23). However, in addition, such a member is also liable to sway or to have end rotations. The new position of the member is shown in the figure with end A moving to A' and end B to B'. The sway v_2 of the second end of the member parallel to Q axis of the member is given by $B'B''$ where from the figure

$$v_2 = B'B'' = df - fe = y_j \cos \alpha - x_j \sin \alpha, \quad (2.26)$$

53

Now
$$-\sin \alpha = \cos \gamma = l_Q, \qquad (2.27)$$

and
$$\cos \alpha = \cos \phi = m_Q, \qquad (2.28)$$

The direction cosines of the Q axis of the member with respect to X and Y axes are l_Q and m_Q respectively. Thus

$$v_2 = l_Q x_j + m_Q y_j \qquad (2.29)$$

similarly
$$v_1 = l_Q x_i + m_Q y_i \qquad (2.30)$$

and the true sway of the member v parallel to Q axis becomes

$$v = v_2 - v_1$$

i.e.

$$
\begin{array}{cc}
\text{at joint } i & \text{at joint } j
\end{array}
$$
$$v = [-l_Q \; -m_Q \ldots l_Q \; m_Q] \qquad \{x_i \; y_i \ldots x_j \; y_j\} \quad (2.31)$$

Because the member is rigidly connected to other members at joints i and j, the rotations θ of its ends are equal to the joint rotations. Thus

$$\theta_{R1} = \theta_i \qquad (2.32)$$

and
$$\theta_{R2} = \theta_j \qquad (2.33)$$

These rotations are in the $X-Y$ plane about R axes of the member.

The joint displacements at each joint are $\{x \; y \; \theta\}$ while the four member displacements are, $\{u \; v \; \theta_{R1} \; \theta_{R2}\}$. These two sets of displacements are, by virtue of equations (2.23), (2.31), (2.32) and (2.33) related as

$$
\begin{bmatrix} u \\ \\ v \\ \\ \theta_{R1} \\ \\ \theta_{R1} \end{bmatrix}
=
\begin{bmatrix}
-l_p & -m_p & 0 & \ldots & l_p & m_p & 0 \\
-l_Q & -m_Q & 0 & \ldots & l_Q & m_Q & 0 \\
0 & 0 & 1 & \ldots & 0 & 0 & 0 \\
0 & 0 & 0 & \ldots & 0 & 0 & 1
\end{bmatrix}
\begin{bmatrix} x_i \\ y_i \\ \theta_i \\ \vdots \\ x_j \\ y_j \\ \theta_j \end{bmatrix}
\quad (2.34)
$$

or
$$\mathbf{Z}_m = [A_1 \quad A_2]\{X_i \quad X_i\} \qquad (2.35)$$

Equations (2.34) and (2.35) express the member displacements for a member m in terms of the displacements for the joints i and j to which it is connected. The member has its first end at joint i

54

and its second end at j. Matrix $A_m = [A_1 \ A_2]$ is called the displacement transformation matrix of the member which has four rows corresponding to the four displacements of the member. The number of columns in either A_1 or A_2 depends on the number of degrees of freedom at joints i or j respectively. The advantage of constructing the displacement transformation matrix in this manner lies in the fact that a member needs to be specified by the joint numbers i and j at its ends. Furthermore it is very easy to devise an automatic method to construct the displacement transformation matrix for every member of the structure by simply inspecting the joint numbers at the ends of each member and then proceeding to form A_1 and A_2 under these joints. Thus if Z is the column vector for the displacements of all the members and X is the column vector for all the joint displacements in the structure, the matrix A in

$$Z = AX \tag{2.36}$$

is the displacement transformation matrix for the whole structure. The number of rows in A is equal to four times the total number

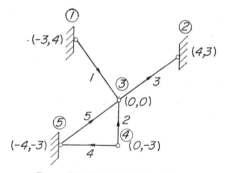

Figure 2.6. Pin-jointed plane frame

of the rigidly jointed members in the structure while the number of its columns is equal to the total degrees of freedom of the structure.

As an example consider the pin-jointed plane frame of Figure 2.6 in which all the joints and the members are numbered at random. The joint numbers are enclosed in circles and the co-ordinates of each joint are given in brackets in the figure. The positive, P, direction of each member is specified by an arrow, the head of which is pointing to the second end of the member. The directions of these arrows are also at random and the member number for each member is placed next to the arrow.

From the figure it is seen that the length of the members are

$$L_1 = 5, L_2 = 3, L_3 = 5, L_4 = 4 \quad \text{and} \quad L_5 = 5.$$

Now for the first member

$$l_{P1} = \frac{0 - (-3)}{5} = 0.6, \quad m_{P1} = \frac{0 - 4}{5} = -0.8.$$

In a similar manner the direction cosines for all the members can be calculated. These are tabulated in Table 2.1.

Table 2.1

Member No.	l_P	m_P
1	0.6	−0.8
2	0	1
3	0.8	0.6
4	−1	0
5	0.8	0.6

Thus the displacement transformation is carried out as follows

$$
\begin{bmatrix} u_1 \\ u_2 \\ u_3 \\ u_4 \\ u_5 \end{bmatrix} =
\begin{bmatrix}
\overset{\text{at joint 3}}{} & & \overset{\text{at joint 4}}{} & \\
0.6 & -0.8 & 0 & 0 \\
0 & 1 & 0 & -1 \\
-0.8 & -0.6 & 0 & 0 \\
0 & 0 & 1 & 0 \\
0.8 & 0.6 & 0 & 0
\end{bmatrix}
\begin{bmatrix} x_3 \\ y_3 \\ x_4 \\ y_4 \end{bmatrix}
$$

A typical member such as member 3 has its first end at joint 3 and therefore $-l_{P3}$ and $-m_{P3}$ are the contributions of this member to the columns of joint 3. This member is not connected to joint 4 and therefore does not contribute to its columns. It is also noticed that the joints, 1, 2, and 5 are excluded from the displacement transformation matrix because, as they are supports, they have no degrees of freedom.

2.2.3. The overall stiffness matrix of a structure

This matrix expresses the equations of joint equilibrium of a structure. It relates the known externally applied forces with the corresponding unknown joint displacements and the solution of these equations, gives the values of the displacements. Suppose that $L = \{L_1 L_2 \ldots\}$ are the applied forces, loads and/or moments, acting on a structure. These forces are vectorially equivalent to the resulting joint displacements $X = \{X_1 X_2 \ldots\}$ That is to say each displacement X_i takes place under the load L_i and in the same direction.

The application of the forces L will induce member forces P in the structure and these members will undergo displacements Z. Since the work done on the structure is equivalent to the strain energy stored in the members, it follows that

$$\tfrac{1}{2}(L_1X_1+L_2X_2+ \ldots) = \tfrac{1}{2}(P_1Z_1+P_2Z_2+ \ldots)$$

but $L_1X_1+L_2X_2+L_3X_3+ \ldots = [L_1L_2L_3 \ldots] \{X_1X_2X_3 \ldots\} = L'X$

and

$$P_1Z_1+P_2Z_2+ \ldots = P'Z$$

$$\therefore\ L'X = P'Z,$$

but from equation (2.36) we have $Z = AX$

$$\therefore\ L'X = P' . AX$$

Now for this equation to hold true for all values of X it follows that

$$L' = P' . A$$

where L' is the transpose of L and P' is the transpose of P. Transposing both sides of these equations:

$$L = A' . P \qquad\qquad (2.37)$$

Equation (2.37) expresses the relationship between the external loads and the member forces, i.e. they are the equilibrium equations.

Earlier on in this chapter the relationship between member forces and dispacements were derived as $P = k . Z$ and substituting for P in equation (2.37) gives $L = A' . k . Z$. Again since $Z = A . X$, it follows that:

$$L = A' . k . A . X \qquad\qquad (2.38)$$

where $A'.k.A = K$ is the required overall stiffness matrix of the structure. The construction of this matrix requires the formation of the matrices k and A derived earlier, although it will be shown later that it is possible to construct K directly.

With the information available so far, it is possible to carry out the analysis of structures, using the matrix displacement method. This can be summarised using the following steps with matrices that have already been derived

(i) Construct the assemblage of member stiffness matrices k in $P = k.Z$.

(ii) Construct the displacement transformation matrix A in $Z = A.X$.

(iii) Construct the overall stiffness matrix K of the structure by the triple multiplication $A'.k.A$.

(iv) Solve the set of equations $L = K.X$ to obtain the unknown joint displacements X.

(v) Calculate the member displacements Z using $Z = A.X$ with A being available from step (ii).

(vi) Calculate the member forces using $P = k.Z$ with k being vailable from step (i).

(vii) Check the externally applied load by using equation (2.37) i.e. $L = A'.P$. with A' being available from step (iii).

Although both k and A are required to calculate the member forces and displacements, it is however, cumbersome to carry out the triple multiplication $A'k.A$ for the construction of the overall stiffness matrix. For this reason it is advantageous to describe the construction of this matrix directly.

(i) For a pin-jointed space structure

The displacement transformation matrix for a member of a pin-jointed space frame is given in equation (2.24) as $[-l_p - m_p - n_p \ldots \ldots l_p \, m_p \, n_p]$. The first three elements are contributed to the columns of the joint (i), to which the first end of the member is connected, and the last three to the joint (j) at the second end of the member. The stiffness of this member is EA/L. Using the notation of (2.9), it follows that

$$kA = [-al_p - am_p - an_p \ldots al_p \, am_p \, an_p]$$

58

Transposing A to A' and post multiplying it by kA gives:

$$
K =
\begin{array}{c}
\text{at} \\ \text{joint} \\ i \\ \\ \text{at} \\ \text{joint} \\ j
\end{array}
\overbrace{
\begin{bmatrix}
al_p^2 & al_p m_p & al_p n_p & \cdots & -al_p^2 & -al_p m_p & -al_p n_p \\
al_p m_p & am_p^2 & am_p n_p & \cdots & -al_p m_p & -am_p^2 & -am_p n_p \\
al_p n_p & an_p m_p & an_p^2 & \cdots & -al_p n_p & -an_p m_p & -an_p^2 \\
\\
-al_p^2 & -al_p m_p & -al_p n_p & \cdots & al_p^2 & al_p m_p & al_p n_p \\
-al_p m_p & -am_p^2 & -am_p n_p & \cdots & al_p m_p & am_p^2 & am_p n_p \\
-al_p n_p & -an_p m_p & -an_p^2 & \cdots & al_p n_p & an_p m_p & an_p^2
\end{bmatrix}}^{\text{at joint } i \qquad\qquad \text{at joint } j}
$$

$$(2.39)$$

This matrix relates the externally applied load vector $L = \{H_i\,V_i\,R_i \ldots H_j\,V_j\,R_j\}$ to the corresponding displacements $X = \{x_i\,y_i\,z_i \ldots \ldots x_j\,y_j\,z_j\}$ at joints i and j. In the expression for the load vector H, V and R are vectorially equivalent to x, y and z.

(ii) For a pin-jointed plane frame

The pin-jointed plane frame is a special case of the pin-jointed space frame with the joints having only x and y degrees of freedom. This means the columns corresponding to z_i and z_j in equation (2.39) no longer exist. Accordingly the rows corresponding to R_i and R_j also disappear. The equation $L = K.X$, therefore is reduced to

$$
\begin{bmatrix} H_i \\ V_i \\ \\ H_j \\ V_j \end{bmatrix}
=
\overbrace{
\begin{bmatrix}
al_p^2 & al_p m_p & \cdots & -al_p^2 & -al_p m_p \\
al_p m_p & am_p^2 & \cdots & -al_p m_p & -am_p^2 \\
\\
-al_p^2 & -al_p m_p & \cdots & al_p^2 & al_p m_p \\
-al_p m_p & -am_p^2 & \cdots & al_p m_p & am_p^2
\end{bmatrix}}^{\text{at joint } i \qquad\qquad \text{at joint } j}
\begin{bmatrix} x_i \\ y_i \\ \\ x_j \\ y_j \end{bmatrix}
\quad (2.40)
$$

(iii) For a rigidly jointed plane frame

Proceeding as in the previous cases and using the matrix k of equation (2.10) and A of equation (2.34). The contribution of a member, linking the joints i and j, to the equations $L = K.X$

becomes:

$$
\begin{bmatrix} H_i \\ V_i \\ M_i \\ \\ H_j \\ V_j \\ M_j \end{bmatrix} =
\begin{bmatrix}
A & B & -C & \cdots & -A & -B & -C \\
B & F & -T & \cdots & -B & -F & -T \\
-C & -T & e & \cdots & C & T & f \\
 & & & & & & \\
-A & -B & C & \cdots & A & B & C \\
-B & -F & T & \cdots & B & F & T \\
-C & -T & f & \cdots & C & T & e
\end{bmatrix}
\begin{bmatrix} x_i \\ y_i \\ \theta_i \\ \\ x_j \\ y_j \\ \theta_j \end{bmatrix} \quad (2.41)
$$

at joint i at joint j

or for simplicity:

$$
\begin{bmatrix} L_i \\ L_j \end{bmatrix} = \begin{bmatrix} K_{ii} & K_{ij} \\ K_{ji} & K_{jj} \end{bmatrix} \begin{bmatrix} X_i \\ X_j \end{bmatrix} \quad (2.41a)
$$

where e and f are defined by equations (2.9) and

$$
\begin{aligned}
A &= al_p^2 + bl_Q^2 \\
B &= al_p m_p + bl_Q m_Q \\
C &= dl_Q \\
F &= am_p^2 + bm_Q^2 \\
T &= dm_Q
\end{aligned} \quad (2.42)
$$

while a, b and d are defined by equations (2.9). The matrices in equations (2.39) through (2.41) are all symmetrical. This follows from Maxwell's Reciprocal theorem.

2.2.4. The use of degrees of freedom in structural representation

In the previous section it was noticed that the degrees of freedom at the joints of a structure decide whether a structure is pin-jointed or rigid, a plane frame or a space structure. In general there are six degrees of fredom at a joint namely x, y, z, θ_X, θ_Y and θ_Z. It is advantageous, from the point of view of presenting the information about a joint to the computer, to represent these degrees of freedom by a six figure integer having 1 for each degree that exists and a zero if it does not. Thus a joint with six degrees of freedom in a space frame will be represented by the integer 111 111. If this joint is restrained in the Y direction, for instance, then it will be represented by the integer 101 111.

In this manner any type of boundary conditions can be specified. In Table 2.2 a list of different types of joints and supports is given together with their integer representations as examples of the scheme and the list can be extended to include every type of joint.

Table 2.2 INTEGER REPRESENTATION SCHEME

Type of Joint	Representation of Degrees of freedom
A joint in a rigid space frame	1 1 1 1 1 1
A joint in a pin-jointed space frame	1 1 1 0 0 0
A joint in a rigid plane frame	1 1 0 0 0 1
A joint in pin-jointed plane frame	1 1 0 0 0 0
A fixed support	0
A roller free in Z direction	1 0 0 0
A roller free in X direction	1 0 0 0 0 0
A pinned support free about X axis	1 0 0
A pinned support free about Z axis	1

These figures can be utilised directly in the construction of the overall stiffness matrix of a structure, as the zeros indicate which of the rows and columns should be omitted. Thus a roller support, free to move along the Z axis will have only one row and one column in the overall stiffness matrix corresponding to the z displacement. This scheme of representation is also advantageous in defining symmetric or antisymmetric behaviour of a frame. For instance the symmetrical pitched roof frame of Figure 2.7 is symmetrically loaded in Figure 2.7(a). For this reason only half the frame, ABC, needs to be analysed. Joint B of this frame can move in X, Y, and θ_z directions and its integer representation is 110 001. On the other hand joint C, which is on the axis of symmetry, only moves vertically, in Y direction, and therefore can be represented by integer 10 000. Similarly in the case of the antisymmetrical deformation of Figure 2.7(b), the integer representation for joint B remains as 110 001, whereas that for joint C becomes 100 001, since it moves in X and θ_z directions.

Furthermore in many cases, the axial deformation due to EA/L effect need not be included in the analysis of frames such as that of Figure 2.7. This means that the movement of joint B in the Y direction can be neglected and thus the integer representation of this joint becomes 100 001.

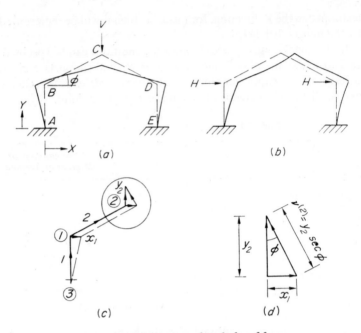

Figure 2.7. Deformations of pitched roof frame
(a) Symmetrical loading
(b) Anti-symmetrical loading
(c) Numbering of joints and members
(d) Vector diagram for displacements

2.2.5. Worked Examples

1. In this example the continuous beam of Figure 2.8 is analysed. The beam is fixed at joint 1 and simply supported at joints 2, 3 and 4. The arrow on each member identifies its first and second ends and defines the direction of the P axis of the member. Thus member 1 has its first end at joint 1 and its second end at joint 2. The length and cross-sectional properties of the members are different and can be identified for each member by the suffixes shown in the figure. The beam is subject to a moment $M_2 = M$. Suppose that it is required to draw the bending moment and shearing force diagram for the beam.

Since joint 1 is fixed it has no degrees of freedom and therefore it is omitted from the overall stiffness matrix. Joints 2, 3 and 4 have

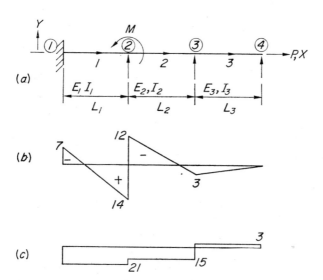

Figure 2.8. Shearing force and bending moment diagram for a continuous beam
(a) Dimensions and loading
(b) B.M. diagram
(c) S.F. diagram

one degree of freedom, namely rotations θ_2, θ_3 and θ_4 respectively, the horizontal movements of these joints being negligible. In this manner the overall stiffness matrix is reduced to a 3×3 matrix, the columns of which correspond to the displacement vector $X = \{\theta_2\ \theta_3\ \theta_4\}$. The rows of this matrix correspond to the load vector $L = \{M_2\ M_3\ M_4\} = \{M\ 0\ 0\}$.

Referring to equations (2.41) for rigidly jointed plane frames, since the rows and columns corresponding to joint translation in X and Y directions are omitted, the contribution of each member to these equations become

$$\begin{matrix} & \text{joint } i \quad \text{joint } j \\ \text{At joint } i \\ \text{At joint } j \end{matrix} \begin{bmatrix} M_i \\ M_j \end{bmatrix} = \begin{bmatrix} e & f \\ f & e \end{bmatrix} \begin{bmatrix} \theta_i \\ \theta_j \end{bmatrix} \qquad (2.43)$$

For member 1, $i = 1$, $j = 2$ and since joint 1 is fixed the contribution of this member to the stiffness matrix is only $e = 4E_1I_1/L_1$ appearing on the row and column corresponding to M_2 and θ_2. For member 2, $i = 2$ and $j = 3$ and thus the contribution of this member will be in accordance with equations (2.43). Likewise for

63

member 3 with $i = 3$ and $j = 4$. The set of equations $\boldsymbol{L} = \boldsymbol{K}.\boldsymbol{X}$ thus becomes

$$
\begin{array}{c}
\text{joint 2} \\ \\ \text{joint 3} \\ \\ \text{joint 4}
\end{array}
\begin{bmatrix} M \\ \\ 0 \\ \\ 0 \end{bmatrix}
=
\overset{\qquad\qquad\text{joint 2}\qquad\qquad\qquad\text{joint 3}\qquad\qquad\qquad\text{joint 4}}{
\begin{bmatrix}
\dfrac{4E_1I_1}{L_1} + \dfrac{4E_2I_2}{L_2} & \dfrac{2E_2I_2}{L_2} & 0 \\
\dfrac{2E_2I_2}{L_2} & \dfrac{4E_2I_2}{L_2} + \dfrac{4E_3I_3}{L_3} & \dfrac{2E_3I_3}{L_3} \\
0 & \dfrac{2E_3I_3}{L_3} & \dfrac{4E_3I_3}{L_3}
\end{bmatrix}}
\begin{bmatrix} \theta_2 \\ \\ \theta_3 \\ \\ \theta_4 \end{bmatrix}
\tag{2.44}
$$

The solution of these equations gives the values of θ_2, θ_3 and θ_4. This completes the steps (i)–(iv) of the matrix displacement method given in Section 2.2.3. In this problem once the joint rotations are found the member end rotations become known and step (v) becomes superfluous. Since $\boldsymbol{Z} = \boldsymbol{A}.\boldsymbol{X}$ and $\boldsymbol{P} = \boldsymbol{k}.\boldsymbol{Z}$, hence $\boldsymbol{P} = \boldsymbol{k}.\boldsymbol{A}.\boldsymbol{X}$ and the member forces are calcuated either by first multiplying $\boldsymbol{k}$ and $\boldsymbol{A}$ or by a direct construction of $\boldsymbol{kA}$. Adopting the latter and using equations (2.10) and (2.34), with the same notations of (2.9) for member forces, these become

$$
\begin{bmatrix} P \\ S \\ M_{R1} \\ M_{R1} \end{bmatrix}
=
\overset{\text{at joint } i\qquad\qquad\qquad\qquad\text{at joint } j}{
\begin{bmatrix}
-al_p & -am_p & 0 & \ldots & al_p & am_p & 0 \\
-bl_Q & -bm_Q & d & \ldots & bl_Q & bm_Q & d \\
-dl_Q & -dm_Q & e & \ldots & dl_Q & dm_Q & f \\
-dl_Q & -dm_Q & f & \ldots & dl_Q & dm_Q & e
\end{bmatrix}}
\begin{bmatrix} x_i \\ y_i \\ \theta_i \\ x_j \\ y_j \\ \theta_j \end{bmatrix}
\tag{2.45}
$$

Equation (2.45) is general and can be used for evaluating the member forces of any rigidly jointed plane frame. In the example under consideration, however, the row corresponding to the axial force p is not required and the columns corresponding to the restraints x_i, y_i, x_j and y_j are removed, thus reducing equations (2.45) to

$$
\begin{bmatrix} S \\ M_{R1} \\ M_{R2} \end{bmatrix}
=
\overset{\text{at } i\ \text{at } j}{
\begin{bmatrix} d & d \\ e & f \\ f & e \end{bmatrix}}
\times
\begin{bmatrix} \theta_i \\ \theta_j \end{bmatrix}
\tag{2.46}
$$

For the members of the beam $P = k.A.X$ become:

$$
\begin{bmatrix} S^{(1)} \\[2ex] M_{R1}^{(1)} \\[2ex] M_{R2}^{(1)} \\[2ex] S^{(2)} \\[2ex] M_{R1}^{(2)} \\[2ex] M_{R2}^{(2)} \\[2ex] S^{(3)} \\[2ex] M_{R1}^{(3)} \\[2ex] M_{R2}^{(3)} \end{bmatrix}
=
\begin{bmatrix}
\overset{\text{joint 2}}{-\dfrac{6E_1I_1}{L_1^2}} & \overset{\text{joint 3}}{0} & \overset{\text{joint 4}}{0} \\[2ex]
\dfrac{2E_1I_1}{L_1} & 0 & 0 \\[2ex]
\dfrac{4E_1I_1}{L_1} & 0 & 0 \\[2ex]
-\dfrac{6E_2I_2}{L_2^2} & -\dfrac{6E_2I_2}{L_2^2} & 0 \\[2ex]
\dfrac{4E_2I_2}{L_2} & \dfrac{2E_2I_2}{L_2} & 0 \\[2ex]
\dfrac{2E_2I_2}{L_2} & \dfrac{4E_2I_2}{L_2} & 0 \\[2ex]
0 & -\dfrac{6E_3I_3}{L_3^2} & -\dfrac{6E_3I_3}{L_3^2} \\[2ex]
0 & \dfrac{4E_3I_3}{L_3} & \dfrac{2E_3I_3}{L_3} \\[2ex]
0 & \dfrac{2E_3I_3}{L_3} & \dfrac{4E_3I_3}{L_3}
\end{bmatrix}
\begin{bmatrix} \theta_2 \\[2ex] \theta_3 \\[2ex] \theta_4 \end{bmatrix}
\qquad (2.47)
$$

The member numbers are given in brackets above each force. For the special case of a continuous beam of constant E and I and with three equal spans L, the solution of the joint equilibrium equations (2.44) gives

$$\{\theta_2 \ \theta_3 \ \theta_4\} = \frac{ML}{52EI}\{7 \ -2 \ 1\}.$$

Substituting these values in equations (2.47) gives the member forces as

$$\frac{M}{26L}\{-21 \ \ 7L \ \ 14L \ \ -15 \ \ 12L \ \ 3L \ \ 3 \ -3L \ \ 0\}$$

The shearing force and the bending moment diagrams can now be drawn. These are shown in Figure 2.8.

2. In this example the joint equilibrium equations $L = K.X$ for the portal of Figure 2.9 will be prepared. It is now obvious that once this task is achieved the rest of the analysis of the frame is straight forward. The members and joints are numbered at random. The arrows indicate that the first end of member 1 and the second end of member 2 are connected to joint 2 while the first end of member 2 and the secnod end of member 3 are connected to joint 3. The arrows also define the P and Q axes of each member and these are shown in the figure.

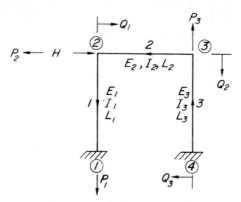

Figure 2.9. Rectangular portal

Once again neglecting the vertical movement of joints 2 and 3, the joint displacement vector reduces to $X = \{x_2 \ \theta_2 \ x_3 \ \theta_3\}$. Corresponding to this, the load vector is $L = \{H_2 \ M_2 \ H_3 \ M_3\} = \{H \ 0 \ 0 \ 0\}$, and the overall stiffness matrix is of the order 4×4. For the first member $i = 2$ and $j = 1$. Since joint 1 is fixed K_{ij}, K_{ji} and K_{jj} submatrices of equations (2.41a) no longer exist for this member, while $K_{ii} = K_{22}$ and the four elements of this submatrix appear in the quadrate corresponding to the rows and columns for joint 2 only. Similarly for the third member, since joint 4 is fixed and $j = 3$ then $K_{jj} = K_{33}$ and its four elements appear in the quadrate for joint 3. For the second member $i = 3$ and $j = 2$, $(i > j)$ and therefore referring to equations (2.41a), K_{ii} appear in the quadrate corresponding to the rows and columns for joint 3. K_{ij} appear in the quadrate that corresponds to the rows of joint 3 and the columns of joint 2, whereas K_{ji} appear on the quadrate corresponding to the rows of joint 2 and the columns of joint 3. Finally K_{jj} appear in the quadrate that belongs to the rows and columns of joint 2. Thus

66

equations (2.41) become

$$
\begin{bmatrix} H \\ 0 \\ 0 \\ 0 \end{bmatrix} =
\begin{bmatrix}
A_1+A_2 & -C_1+C_2 & -A_2 & C_2 \\
-C_1+C_2 & e_1+e_2 & -C_2 & f_2 \\
-A_2 & -C_2 & A_2+A_3 & -C_2+C_3 \\
C_2 & f_2 & -C_2+C_3 & e_2+e_3
\end{bmatrix}
\begin{bmatrix} x_2 \\ \theta_2 \\ x_3 \\ \theta_3 \end{bmatrix}
$$

$$\text{joint 2} \qquad\qquad\qquad \text{joint 3}$$

where A and C are defined by equations (2.42). Because the members of this frame are orthogonal and P, Q axes of the members are along X, Y axes of the overall system coordinates, the direction cosines of the members are either zeroes or unity. Here $l_{p1} = l_{Q2} = l_{p3} = 0$, $l_{Q1} = -l_{p2} = -l_{Q3} = 1$. Using these and equations (2.42). $L = K \cdot X$ becomes

$$
\begin{bmatrix} H \\ 0 \\ 0 \\ 0 \end{bmatrix} =
\begin{bmatrix}
b_1+a_2 & -d_1 & -a_2 & 0 \\
-d_1 & e_1+e_2 & 0 & f_2 \\
-a_2 & 0 & a_2+b_3 & -d_3 \\
0 & f_2 & -d_3 & e_2+e_3
\end{bmatrix}
\begin{bmatrix} x_2 \\ \theta_2 \\ x_3 \\ \theta_3 \end{bmatrix} \quad (2.48)
$$

It is useful to note that for the special case of the antisymmetrical deformation, when $E_1 = E_3$, $I_1 = I_3$ and $L_1 = L_3$, the resulting displacements $x_2 = x_3$ and $\theta_2 = \theta_3$. In this case, when analysing the whole frame it is of advantage to make use of this and construct the 2×2 overall stiffness matrix to correspond to the displacements $X = \{x_2\ \theta_2\}$ and $L = \{H\ O\}$. In this matrix rows and columns corresponding to joint 3 do not appear and the contributions of the members to these rows and columns go to those of joint 2. This simply amounts to the matrix addition of the four submatrices of equations (2.48) giving:

$$
\begin{bmatrix} H \\ 0 \end{bmatrix} =
\begin{bmatrix}
b_1+b_3 & -d_1-d_3 \\
-d_1-d_3 & e_1+2e_2+2f_2+e_3
\end{bmatrix}
\begin{bmatrix} x_2 \\ \theta_2 \end{bmatrix} \quad (2.49)
$$

This procedure is better visualised by constructing the overall stiffness matrix using the triple multiplication $A' \cdot k \cdot A$. The matrix k for the whole structure will be similar to that shown in equation (2.11) with $n = 3$ for three members. With $X = \{x_2\ \theta_2\}$ the displace-

6*

ment transformation matrix A becomes

$$
\begin{bmatrix}
v^{(1)} \\
\theta^{(1)}_{R1} \\
\theta^{(1)}_{R2} \\
v^{(2)} \\
\theta^{(2)}_{R1} \\
\theta^{(2)}_{R2} \\
v^{(3)} \\
\theta^{(3)}_{R1} \\
\theta^{(3)}_{R2}
\end{bmatrix}
=
\begin{bmatrix}
-1 & 0 \\
0 & 1 \\
0 & 0 \\
0 & 0 \\
0 & 1 \\
0 & 1 \\
-1 & 0 \\
0 & 0 \\
0 & 1
\end{bmatrix}
\begin{bmatrix}
x_2 \\
\theta_2
\end{bmatrix}
,
$$

which is used with k in the triple multiplication leading to the same result as equation (2.49)

3. As a final example the joint equilibrium equation for the symmetrical case of the pitched roof frame of Figure 2.7(a) is constructed. Because of symmetry only half the frame with half the load at the apex needs to be considered. For this case joint 2 only moves vertically while $x_2 = \theta_2 = 0$. On the other hand only two degrees of freedom at joint 1 are significant, namely x_1 and θ_1, the numbering of joints and member are shown in Figure 2.7(c). This reduces the joint displacement vector to $X = \{x_1 \, \theta_1 \, y_2\}$. However considering the vector diagram for displacements Figure 2.7(d), it is evident that $x_1 = y_2 \tan \phi$ and the sway of member 2 parallel to its Q axis is $v^{(2)} = y_2 \sec \phi$, where ϕ is the angle of pitch for the frame. It is therefore possible to reduce the joint displacement vector to only two elements, i.e. $X = \{\theta_1 \, y_2\}$. Terms involving x_1 require to be multiplied by $\tan \phi$ and expressed in terms of y_2. For instance the sway of member 1 parallel to its Q axis is $v^{(1)} = -x_1 = -y_2 \tan \phi$. The displacement transformation matrix can now be formed

$$
\begin{bmatrix}
v^{(1)} \\
\theta^{(1)}_{R1} \\
\theta^{(1)}_{R2} \\
v^{(2)} \\
\theta^{(2)}_{R1} \\
\theta^{(2)}_{R2}
\end{bmatrix}
=
\begin{bmatrix}
0 & -\tan \phi \\
0 & 0 \\
1 & 0 \\
0 & \sec \phi \\
1 & 0 \\
0 & 0
\end{bmatrix}
\begin{bmatrix}
\theta_1 \\
y_2
\end{bmatrix}
$$

and by the usual procedure and using the notations of (2.9) and (2.42) $L = K \cdot X = A' \cdot k \cdot A \cdot X$ becomes

$$
\begin{bmatrix} 0 \\ \dfrac{-V}{2} \end{bmatrix} = \begin{bmatrix} e_1 + e_2 & -d_1 \tan \phi + d_2 \sec \phi \\ -d_1 \tan \phi + d_2 \sec \phi & b_1 \tan^2 \phi + b_2 \sec^2 \phi \end{bmatrix} \begin{bmatrix} \theta_1 \\ y_2 \end{bmatrix}
$$

The reader may prepare these equations directly using the procedure of (2.41).

2.3. MATRIX FORCE METHOD

2.3.1. The flexibility matrix

(i) For pin-jointed structures

For a member of a pin-jointed structure such as that of section 2.2. (i) and shown in Figure 2.1, equation (2.1), was obtained using Hooke's law to be

$$
p = \frac{EA}{L} u
$$

This equation can be rearranged so that the extension of the member is expressed in terms of the force; thus

$$
u = \frac{L}{EA} p \tag{2.50}
$$

Here L/EA is the flexibility of the member. For all the n members of a structure similar equations can be written. In matrix form these become

$$
\begin{bmatrix} u_1 \\ u_2 \\ \vdots \\ u_i \\ \vdots \\ u_n \end{bmatrix} = \begin{bmatrix} L_1/E_1A_1 & & & & & 0 \\ & L_2/E_2A_2 & & & & \\ & & \ddots & & & \\ & & & L_i/E_iA_i & & \\ & & & & \ddots & \\ 0 & & & & & L_n/E_nA_n \end{bmatrix} \begin{bmatrix} p_1 \\ p_2 \\ \vdots \\ p_i \\ \vdots \\ p_n \end{bmatrix}
$$

$$
\tag{2.51}
$$

69

or simply $Z = fP$

where f is the flexibility matrix of the members which is again diagonal with each element expressing the flexibility of a member.

(ii) For rigidly-jointed plane frames

The member AB of Figure 2.10 is under pure rotation due to the action of the end moments M_{AB} and M_{BA}. The slope deflection equations for this member are

$$M_{AB} = \frac{4EI}{L}\theta_{AB} + \frac{2EI}{L}\theta_{BA},$$

$$M_{BA} = \frac{2EI}{L}\theta_{AB} + \frac{4EI}{L}\theta_{BA}.$$

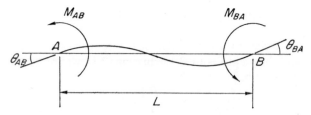

Figure 2.10. Pure rotation of a member

Solving these for θ_{AB} and θ_{BA}

$$\theta_{AB} = \frac{L}{3EI}M_{AB} - \frac{L}{6EI}M_{BA}$$

$$\theta_{BA} = -\frac{L}{6EI}M_{AB} + \frac{L}{3EI}M_{BA},$$

and in Matrix form

$$\begin{bmatrix} \theta_{AB} \\ \theta_{BA} \end{bmatrix} = \begin{bmatrix} \dfrac{L}{3EI} & \dfrac{-L}{6EI} \\ \dfrac{-L}{6EI} & \dfrac{L}{3EI} \end{bmatrix} \begin{bmatrix} M_{AB} \\ M_{BA} \end{bmatrix} = \frac{L}{6EI} \begin{bmatrix} 2 & -1 \\ -1 & 2 \end{bmatrix} \begin{bmatrix} M_{AB} \\ M_{BA} \end{bmatrix}$$

(2.52)

or just $\boldsymbol{\theta}_i = f_i M_i$ (2.53)

where $\boldsymbol{\theta}_i = \{\theta_{AB}\ \theta_{BA}\}_i$ for member i, $M_i = \{M_{AB}\ M_{BA}\}_i$ and f_i is the flexibility matrix of the member in bending.

For all the n members of a frame equations similar to (2.52) and (2.53) can be constructed and compounded so that they have a common leading diagonal, resulting in

$$
\begin{bmatrix} \theta_1 \\ \theta_2 \\ \cdot \\ \theta_i \\ \cdot \\ \theta_n \end{bmatrix} = \begin{bmatrix} f_1 & & & & O \\ & f_2 & & & \\ & & \cdot & & \\ & & & f_i & \\ & & & & \cdot \\ O & & & & f_n \end{bmatrix} \begin{bmatrix} M_1 \\ M_2 \\ \cdot \\ M_i \\ \cdot \\ M_n \end{bmatrix} \qquad (2.54)
$$

i.e. $Z = fP$. $\qquad$ (2.55)

In equtions (2.54) and (2.55) the vector

$$ Z = \{\theta_{1,1}\ \theta_{2,1}\ \theta_{1,2}\ \theta_{2,2}\ \ldots\ \theta_{1,n}\ \theta_{2,n}\} $$

expresses all end rotations of the members. Here the subscripts 1 and 2 refer to the first and second ends of the member respectively, while the subscript (n) refers to the member number. Similarly the vector P summarises the end moments for all the members. Finally matrix f is an assemblage of the flexibility submatrices of the members.

2.3.2. The overall flexibility matrix of a structure

The equilibrium conditions between the externally applied loads L on a structure and the resulting member forces were expressed in equation (2.37) as $L = A'.P$. Unless the forces P are known, however, equations (2.37) can not be easily utilised. Usually the external forces are known and the resulting member forces are required. It is therefore advantageous to express the latter in terms of the former. This is done as

$$
\begin{bmatrix} P_1 \\ P_2 \\ \cdot \\ P_i \\ \cdot \\ P_n \end{bmatrix} = \begin{bmatrix} B_{11} & B_{12} & \ldots & B_{1m} \\ B_{21} & B_{22} & \ldots & B_{2m} \\ & & & \\ & & & \\ & & & \\ B_{n1} & B_{n2} & \ldots & B_{nm} \end{bmatrix} \begin{bmatrix} L_1 \\ L_2 \\ \cdot \\ L_j \\ \cdot \\ L_m \end{bmatrix} \qquad (2.56)
$$

71

i.e. $P = B.L.$ (2.57)

Matrix B is called the load transformation matrix and can be constructed from the equilibrium conditions in the structure.

It has already been shown, in Section 2.2.3 that $L'.X = P'.Z$ and since from equation (2.57) $P = B.L$, i.e. $P' = L'.B'$, it follows that

$$L'.X = L'.B'.Z \qquad (2.58)$$

and for equation (2.58) to hold under any system of load L

$$X = B'.Z \qquad (2.59)$$

Further since $Z = f.P$, therefore $X = B'.f.P$. Again using $P = B.L$, equations (2.59) become

$$X = B'.f.B.L = F.L \qquad (2.60)$$

where $F = B'.f.B$ is the overall flexibility matrix of the structure. Comparing this matrix with the overall stiffness matrix of the structure, K which also relates the external loads and the resulting joint displacement as $L = K.X$, it is observed that $F = K^{-1}$. In other words the flexibility matrix is the inverse of the stiffness matrix and hence it is also square and symmetrical. The duality between the matrices f, B and F and k, A and K can be readily recognised.

2.3.3. Matrix force method for the statically determinate structures

In the case of statically determinate structures, the equilibrium equations $P = B.L$ can be readily constructed by statics and then used in equation (2.60) to calculate the displacements of the structure under the applied loads. Furthermore, the process can be simplified by omitting the construction of the overall flexibility matrix. The resulting steps for the matrix force method are as follows:

(i) Construct the load transformation matrix B and calculate P using $P = B.L$.

(ii) From $Z = f.P$ calculate the member displacements Z and, finally

72

(iii) Calculate the displacements X under the external loads using equations (2.59) i.e. $X = B'.Z$.

For instance the statically determinate frame of constant cross-section in Figure 2.11 is subject at C to a vertical load V and a moment M as shown. In order to construct the load transformation matrix it is necessary to use the same sign convention for the member forces as that used in deriving the flexibility matrix f and shown in Figure 2.10. For the column AB these moments are shown

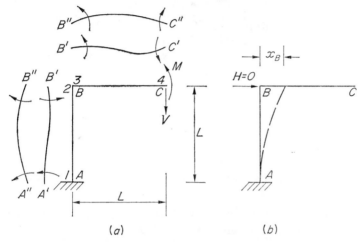

Figure 2.11. Deformation of a rigid-statically determinate frame
(a) Frame and loading
(b) Column sway

as $A'B'$ drawn next to AB in Figure 2.11(a). The applied force V, however bends the column in single curvature as shown by $A''B''$ which indicates that while moment at A due to V is of the same positive sign as that shown at A', the moment at B is of the opposite sign to that shown by B' and therefore is negative.

The bending moments at these points due to the external moment would, of course, have opposite signs to those due to V. Taking moments about point A, marked 1 in the figure, the bending moment there is

$$M_1 = LV - M$$

Similarly the bending moments at the other points 2, 3 and 4 are

73

found. In matrix form these are given as

$$
P = \begin{bmatrix} M_1 \\ M_2 \\ M_3 \\ M_4 \end{bmatrix} = \begin{bmatrix} L & -1 \\ -L & 1 \\ L & -1 \\ 0 & 1 \end{bmatrix} \cdot \begin{bmatrix} V \\ M \end{bmatrix} = \begin{bmatrix} LV - M \\ -LV + M \\ LV - M \\ M \end{bmatrix}
$$

The corresponding member rotations can now be calculated from

$$
\begin{bmatrix} \theta_1 \\ \theta_2 \\ \theta_3 \\ \theta_4 \end{bmatrix} = \frac{L}{6EI} \begin{bmatrix} 2 & -1 & 0 & 0 \\ -1 & 2 & 0 & 0 \\ 0 & 0 & 2 & -1 \\ 0 & 0 & -1 & 2 \end{bmatrix} \begin{bmatrix} LV - M \\ -LV + M \\ LV - M \\ M \end{bmatrix} =
$$

$$
= \frac{L}{6EI} \begin{bmatrix} 3LV - 3M \\ 3M - 3LV \\ 2LV - 3M \\ 3M - LV \end{bmatrix}
$$

Finally the vertical displacement and rotation $\{y_c\; \theta_c\}$ at C under the external forces are given by $X = B'Z$, viz.

$$
\begin{bmatrix} y_c \\ \theta_c \end{bmatrix} = \begin{bmatrix} L & -L & L & 0 \\ -1 & 1 & -1 & 1 \end{bmatrix} \frac{L}{6EI} \begin{bmatrix} 3LV - 3M \\ 3M - 3LV \\ 2LV - 3M \\ 3M - LV \end{bmatrix}
$$

Hence

$$
y_c = \frac{8L^3V - 9ML^2}{6EI}
$$

in the same direction as V and

$$
\theta_c = \frac{2ML - 1 \cdot 5L^2V}{EI},
$$

in the same direction as M.

It is noticed that only the displacements under the applied external loads are obtained. It is easy, however, to obtain the displacement at any other point by simply applying a force there and remembering that its magnitude is zero. This only adds an extra column to the force transformation matrix. Thus to obtain the sway x_B of the column a horizontal load $H = 0$ is applied at B as shown in Figure 2.11(b). The extra column in the Matrix B will thus be $\{L\ O\ O\ O\}$. This alters the first element of P from $LV - M$ to $LV - M + LH$. Vector Z changes accordingly to

$$
\begin{bmatrix} \theta_1 \\ \theta_2 \\ \theta_3 \\ \theta_4 \end{bmatrix} = \frac{L}{6EI} \begin{bmatrix} 3LV - 3M + 2HL \\ -3LV + 3M - LH \\ 2LV - 3M \\ 3M - LV \end{bmatrix}
$$

and upon premultiplication by the transpose of the extra column of B, i.e. by $[L\ O\ O\ O]$, this gives:

$$
x_B = \frac{L^2}{6EI} (3LV - 3M + 2LH).
$$

Noting that $H = O$, x_B reduces to

$$
x_B = \frac{L^3 V - ML^2}{2EI}
$$

which is the required sway.

2.3.4. Matrix force method for statically indeterminate structure

In a statically indeterminate structure the applied load system may be divided into two categories. These are the external loads L_b acting on the basic statically determinate structure and a set of redundant reactions L_r. The load vector is therefore $L = \{L_b\ L_r\}$. Likewise, the corresponding displacements would be $X = \{X_b\ X_r\}$. Where X_b are the displacements under the loads L_b and X_r are the displacements corresponding to L_r. Generally the latter displacements are known. For instance, the movement of support reactions may be either zeros or fixed quantities.

The overall flexibility matrix can be partitioned into submatrices so that each set of loads may be associated with the vectorially equivalent displacements. Thus

$$\begin{bmatrix} X_b \\ X_r \end{bmatrix} = \begin{bmatrix} F_{bb} & F_{br} \\ F_{rb} & F_{rr} \end{bmatrix} \begin{bmatrix} L_b \\ L_r \end{bmatrix} \tag{2.61}$$

Writing equations (2.61) as two separate sets

$$X_b = F_{bb} \quad L_b + F_{br} \quad L_r \tag{2.62}$$

$$X_r = F_{rb} \quad L_b + F_{rr} \quad L_r \tag{2.63}$$

and from the last set

$$F_{rr} \, L_r = X_r - F_{rb} \, L_b$$
$$L_r = F_{rr}^{-1} . (X_r - F_{rb} . L_b) \tag{2.64}$$

which gives the values of the redundant forces L_r.

The load transformation matrix B can also be partitioned by columns so that the equilibrium equations become

$$P = BL = [B_b \quad B_r] \{L_b \quad L_r\} = B_b \, L_b + B_r \, L_r \tag{2.65}$$

Furthermore this process of partitioning can be used with the triple product $B' f B$, for

$$F = B' \, f \, B = \begin{bmatrix} B_b' \\ \hline B_r' \end{bmatrix} . f . [B_b \mid B_r] = \begin{bmatrix} B_b' \, f \, B_b & B_b' \, f \, B_r \\ \hline B_r' \, f \, B_b & B_r' \, f \, B_r \end{bmatrix} \tag{2.66}$$

Comparing the submatrices of the overall flexibility matrix as given in (2.61) and (2.66), it follows

$$\begin{bmatrix} F_{bb} & F_{br} \\ F_{rb} & F_{rr} \end{bmatrix} = \begin{bmatrix} B_b' \, f \, B_b & B_b' \, f \, B_r \\ B_r' \, f \, B_b & B_r' \, f \, B_r \end{bmatrix}$$

That is:

$$\left. \begin{array}{l} F_{bb} = B_b' \, f \, B_b \\ F_{br} = B_b' \, f \, B_r \\ F_{rb} \quad B_r' \, f \, B_b \\ F_{rr} \quad B_r' \, f \, B_r \end{array} \right\} \tag{2.67}$$

The redundant forces L_r of equations (2.64) will thus be given by

$$L_r = (B_r' \, f \, B_r)^{-1} . (X_r - B_r' \, f \, B_b \, L_b). \tag{2.68}$$

The required steps for the analysis of statically indeterminate structures, using the matrix force method, may be summarized as follows

76

(i) Construct $B'_r \, f \, B_r$ and $B'_r \, f \, B_b \, L_b$.

(ii) Using the known values of X_r calculate the values of the redundants L_r with the aid of equations (2.68).

(iii) Calculate the member forces P using equation (2.65).

(iv) Finally if the displacements X_b are required these may be obtained from equation (2.62).

It is evident that the whole analysis requires the construction of three matrices f, B_b and B_r. The construction of the member flexibility matrix f has been amply dealt with in Section 2.3.1, equations (2.51) through (2.55). The construction of B_b and B_r is no more difficult than that given in Section 2.3.2 and produced for the frame of Figure 2.11. The only difference lies in the partitioning by columns of B into two. The first submatrix corresponds to the basic statically determinate structure and the second to the redundants which are treated as external loads. This is conveniently performed by removing the redundants and constructing B_b for the resulting basic structure under the applied loads. Matrix B_r is then constructed by removing the applied loads and reimposing the redundants.

2.3.5. Worked examples

First of all consider the statically indeterminate pin-jointed structure of Figure 2.12. In order to derive the equilibrium equations between the externally applied load H and the resulting member forces in the basic structure, the redundant members 2 and 4 are removed as shown in Figure 2.12(b). Resolving the remaining forces horizontally and then vertically at C, gives

$$p_1 = -p_3 = H/\sqrt{2}.$$

Writing this in matrix form and including the values of the redundants $p_2 = p_4 = 0$ in the member force vector

$$\begin{bmatrix} p_1 \\ p_2 \\ p_3 \\ p_4 \end{bmatrix} = \begin{bmatrix} 1/\sqrt{2} \\ 0 \\ -1/\sqrt{2} \\ 0 \end{bmatrix} [H], \qquad B_b = \begin{bmatrix} 1/\sqrt{2} \\ 0 \\ -1/\sqrt{2} \\ 0 \end{bmatrix}$$

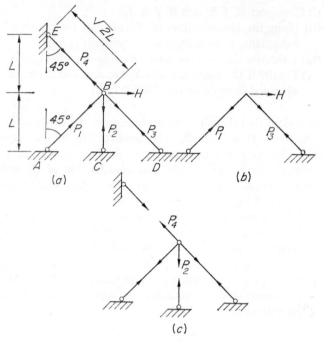

Figure 2.12. The forces in a statically indeterminate pin-jointed structure
(a) Structure and loading
(b) The basic structure
(c) The redundants

Next the external load H is removed and the basic structure is subjected to the redundant forces only as shown in Fig. 2.12(c). Resolving the forces again in the vertical and horizontal directions gives

$$p_1 = -p_2/\sqrt{2} \quad \text{and} \quad p_3 = p_4 - p_2/\sqrt{2}.$$

Again writing these in matrix form:

$$\begin{bmatrix} p_1 \\ p_2 \\ p_3 \\ p_4 \end{bmatrix} = \begin{bmatrix} -1/\sqrt{2} & 0 \\ 1 & 0 \\ -1/\sqrt{2} & 1 \\ 0 & 1 \end{bmatrix} \begin{bmatrix} p_2 \\ p_4 \end{bmatrix} \qquad B_r = \begin{bmatrix} -1/\sqrt{2} & 0 \\ 1 & 0 \\ -1/\sqrt{2} & 1 \\ 0 & 1 \end{bmatrix}$$

The member flexibility matrix f can be constructed directly, using

78

(2.51) with $n = 4$, thus

$$f = \frac{L}{EA}\begin{bmatrix} \sqrt{2} & 0 & 0 & 0 \\ 0 & 1 & 0 & 0 \\ 0 & 0 & \sqrt{2} & 0 \\ 0 & 0 & 0 & \sqrt{2} \end{bmatrix}$$

Hence $B_r' f B_r = \begin{bmatrix} -1/\sqrt{2} & 1 & -1/\sqrt{2} & 0 \\ 0 & 0 & 1 & 1 \end{bmatrix}$.

$$\cdot \frac{L}{EA}\begin{bmatrix} \sqrt{2} & 0 & 0 & 0 \\ 0 & 1 & 0 & 0 \\ 0 & 0 & \sqrt{2} & 0 \\ 0 & 0 & 0 & \sqrt{2} \end{bmatrix}\begin{bmatrix} -1/\sqrt{2} & 0 \\ 1 & 0 \\ -1/\sqrt{2} & 1 \\ 0 & 1 \end{bmatrix}$$

i.e. $\qquad B_r' f B_r = \dfrac{L}{EA}\begin{bmatrix} \sqrt{2}+1 & -1 \\ -1 & 2\sqrt{2} \end{bmatrix}$,

$$B_r' f B_b = \frac{L}{EA}\cdot\begin{bmatrix} -1 & 1 & -1 & 0 \\ 0 & 0 & \sqrt{2} & \sqrt{2} \end{bmatrix}\cdot\begin{bmatrix} 1/\sqrt{2} \\ 0 \\ -1/\sqrt{2} \\ 0 \end{bmatrix} = \frac{L}{EA}\cdot\begin{bmatrix} 0 \\ -1 \end{bmatrix}$$

where $\dfrac{L}{EA}\cdot\begin{bmatrix} -1 & 1 & -1 & 0 \\ 1 & 0 & \sqrt{2} & \sqrt{2} \end{bmatrix}$ is the product $B_r' f$ constructed during the formation of $B_r' f B_r$.

Similarly $B_b' f B_r' = L/EA.[0 \;\; -1]$ which we notice to be the transpose of $B_r' f B_b$. Finally $B_b' f B_b = L\sqrt{2}/EA$, i.e. just a number which is a 1×1 matrix.

In this problem the forces in the members 2 and 4 were taken as internal redundants. The displacement X_r under these forces would be an initial lack of fit in these members. Accordingly, in this case, the vector $X_r' = \{0 \;\; 0\}$. Substituting these together with $B_r' f B_r$, $B_r' f B_b$ and $L_b = H$ into equations (2.68) gives

$$\begin{bmatrix} p_2 \\ p_4 \end{bmatrix} = \left(\frac{L}{EA}\cdot\begin{bmatrix} \sqrt{2}+1 & -1 \\ -1 & 2\sqrt{2} \end{bmatrix}\right)^{-1}\left\{\begin{bmatrix} 0 \\ 0 \end{bmatrix} - \frac{L}{EA}\cdot\begin{bmatrix} 0 \\ -1 \end{bmatrix}[H]\right\}$$

i.e.

$$\begin{bmatrix} p_2 \\ p_4 \end{bmatrix} = \left(\frac{L}{EA} \cdot \begin{bmatrix} \sqrt{2}+1 & -1 \\ -1 & 2\sqrt{2} \end{bmatrix} \right)^{-1} \begin{bmatrix} 0 \\ \dfrac{LH}{EA} \end{bmatrix}$$

Solving these $p_2 = \dfrac{H}{3+2\sqrt{2}}$, $p_4 = \dfrac{(1+\sqrt{2})H}{3+2\cdot\sqrt{2}}$.

Using equations (2.65) which gives $P = B_b L_b + B_r L_r$ results in

$$\dot{P}_1 = \frac{2H(1+\sqrt{})}{3\sqrt{2}+4} \quad \text{and} \quad p_3 = \frac{-H(1+\sqrt{})}{3+2\sqrt{2}}.$$

Finally using equations (2.62) for the horizontal displacement X_b

$$X_b = F_{bb}\, L_b + F_{br}\, L_r =$$
$$= \frac{L\sqrt{2}}{EA} \cdot H + \frac{L}{EA} \cdot [0 \;\; -1] \left\{ \frac{H}{3+\sqrt{2}} \quad \frac{H(1+\sqrt{})}{3+2\sqrt{2}} \right\}$$

$$\therefore \; X_b = HL/EA.$$

It is clear by now that the main task is the construction of the matrices B_b and B_r. As a second example these matrices are constructed for the continuous beam of Figure 2.13. The reactions

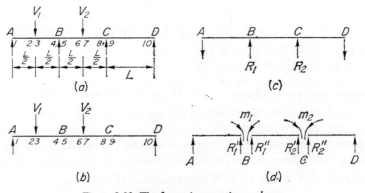

Figure 2.13. The forces in a continuous beam

(a) Beam and loading
(b) Basic structure
(c) External redundants
(d) Internal redundants

at B and C are selected as external redundants and in order to construct B_b they are first removed leaving the basic structure of Figure 2.13(b). The points where the bending moments are calculated in this simply supported beam are numbered from 1 to 10. These are under the applied loads and at the supports. If the load V_1 acts alone the reaction $R_A = 5V_1/6$ and the bending moments at the required points are

$$M_1 = 0, \qquad\qquad M_2 = \frac{5V_1}{6} \cdot \frac{L}{2} = \frac{5LV_1}{12},$$

$$M_3 = \frac{-5LV_1}{12}, \qquad\qquad M_4 = \frac{5V_1}{6} \cdot L - V_1 \cdot \frac{L}{2} = \frac{V_1 L}{3},$$

$$M_5 = \frac{-5V_1}{6}L + V_1 \cdot \frac{L}{2} = \frac{-V_1 L}{3}, \quad M_6 = \frac{5V_1}{6} \cdot \frac{3L}{2} - V_1 L = \frac{V_1 L}{4},$$

$$M_7 = \frac{-V_1 L}{4} \qquad\qquad M_8 = \frac{5V_1}{6} \cdot 2L - V_1 \cdot \frac{3L}{2} = \frac{V_1 L}{6},$$

$$M_9 = \frac{-V_1 L}{6} \qquad\qquad \text{and} \quad M_{10} = 0.$$

Similarly for V_2 acting alone $R_A = V_2/2$ and the bending moments M_1 to M_{10} are calculated.

In matrix form these bending moments are

$$
\begin{bmatrix} M_1 \\ M_2 \\ M_3 \\ M_4 \\ M_5 \\ M_6 \\ M_7 \\ M_8 \\ M_9 \\ M_{10} \end{bmatrix} =
\begin{bmatrix} 0 & 0 \\ 5L/12 & L/4 \\ -5L/12 & -L/4 \\ L/3 & L/2 \\ -L/3 & -L/2 \\ L/4 & 3L/4 \\ -L/4 & -3L/4 \\ L/6 & L/2 \\ -L/6 & -L/2 \\ 0 & 0 \end{bmatrix}
\begin{bmatrix} V_1 \\ V_2 \end{bmatrix} \quad B_b =
\begin{bmatrix} 0 & 0 \\ 5L/12 & L/4 \\ -5L/12 & -L/4 \\ L/3 & L/2 \\ -L/3 & -L/2 \\ L/4 & 3L/4 \\ -L/4 & -3L/4 \\ L/6 & L/2 \\ -L/6 & -L/2 \\ 0 & 0 \end{bmatrix}
$$

The loads V_1 and V_2 are then removed and the basic simply supported beam is subjected to two forces R_1 and R_2 acting upwards at B and C as indicated at Figure 2.13(c). Once again due to the action of R_1 alone, the reaction at A is $R_A = 2R_1/3$ acting down and the bending moments at points 1 to 10 are

$$M_1 = 0, \quad M_2 = \frac{-2R_1}{3} \cdot \frac{L}{2} = \frac{-LR_1}{3}, \quad M_3 = \frac{LR_1}{3},$$

$$M_4 = \frac{-2R_1}{3} \cdot L = \frac{-2LR_1}{3}$$

and so on. These moments are then calculated, with R_2 acting alone and the two sets are listed in matrix form as

$$
\begin{bmatrix} M_1 \\ M_2 \\ M_3 \\ M_4 \\ M_5 \\ M_6 \\ M_7 \\ M_8 \\ M_9 \\ M_{10} \end{bmatrix}
=
\begin{bmatrix}
0 & 0 \\
-L/3 & -L/6 \\
L/3 & L/6 \\
-2L/3 & -L/3 \\
2L/3 & L/3 \\
-L/2 & -L/2 \\
L/2 & L/2 \\
-L/3 & -2L/3 \\
L/3 & 2L/3 \\
0 & 0
\end{bmatrix}
\begin{bmatrix} R_1 \\ R_2 \end{bmatrix}
\qquad
B_r =
\begin{bmatrix}
0 & 0 \\
-L/3 & -L/6 \\
L/3 & L/6 \\
-2L/3 & -L/3 \\
2L/3 & L/3 \\
-L/2 & -L/2 \\
L/2 & L/2 \\
-L/3 & -2L/3 \\
L/3 & 2L/3 \\
0 & 0
\end{bmatrix}
$$

In Figure 2.13(d) the internal moments m_1 and m_2 are shown as an alternative set of redundants. These can be evaluated in terms of the components R_1', R_1'', R_2' and R_2'' of the external redundants by considering the equilibrium of the two systems, i.e. For AB, taking moments about A

$$R_1'L = m_1 \quad \therefore \quad R_1' = m_1/L.$$

For BC, taking moments about C

$$R_1''L = m_1 - m_2 \quad \therefore \quad R_1'' = (m_1 - m_2)/L$$

Hence $R_1 = R_1' + R_1'' = (2m_1 - m_2)/L$ \hfill (2.69)

Similarly

$$R_2 = (2m_2 - m_1)/L \hfill (2.70)$$

Writing equations (2.69) and (2.70) in matrix form

$$\begin{bmatrix} R_1 \\ R_2 \end{bmatrix} = \begin{bmatrix} 2/L & -1/L \\ -1/L & 2/L \end{bmatrix} \begin{bmatrix} m_1 \\ m_2 \end{bmatrix} \qquad (2.71$$

In equations (2.71) the external redundants are $L_{re} = \{R_1 \; R_2\}$, while the internal redundants are $L_{ri} = \{m_1 \; m_2\}$.
Thus equation (2.71) can be written as

$$L_{re} = B_{ie} \cdot L_{ri}, \qquad (2.72)$$

where B_{ie} is the connecting matrix expressing the equilibrium conditions between the two sets of redundancies.

Furthermore it can be proved that

$$B_{ri} = B_{re} \cdot B_{ie} \qquad (2.73)$$

where suffixes i and e refer to internal and external redundants respectively. Although the example of this continuous beam was used to derive these equations, it should be accepted that equations (2.72) and (2.73) are perfectly general. These equations are useful in relating one set of redundancies with another and once the matrix B_r is obtained for one set of redundants, equation (2.73) can be employed to construct directly the matrix B_r for another set of redundants. Furthermore, this equation can be used with any two sets of redundants irrespective of their nature.

For the continuous beam under consideration the new matrix B_{ri} can now be constructed

$$B_{ri} = \begin{bmatrix} 0 & 0 \\ -L/3 & -L/6 \\ L/3 & L/6 \\ -2L/3 & -L/3 \\ 2L/3 & L/3 \\ -L/2 & -L/2 \\ L/2 & L/2 \\ -L/3 & -2L/3 \\ L/3 & 2L/3 \\ 0 & 0 \end{bmatrix} \begin{bmatrix} 2/L & -1/L \\ -1/L & 2/L \end{bmatrix} = \begin{bmatrix} 0 & 0 \\ -1/2 & 0 \\ 1/2 & 0 \\ -1 & 0 \\ 1 & 0 \\ -1/2 & -1/2 \\ 1/2 & 1/2 \\ 0 & -1 \\ 0 & 1 \\ 0 & 0 \end{bmatrix}$$

$$(2.74)$$

Finally matrix B_r for the portal of Figure 2.14 is constructed taking the internal moments M_B, M_C and M_D as the redundants, as shown in Figure 2.14(b). Considering the effect of M_B alone

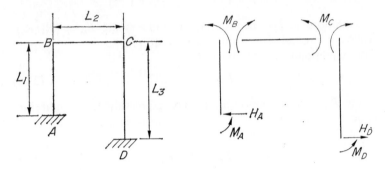

Figure 2.14. The forces in a portal
(a) Dimensions
(b) Redundants and member forces

and taking moments about point C for the member $CD : H_D.L_3 - -M_C+M_D = 0$,

and for
$$M_C = M_D = 0; \quad H_D = 0.$$

From the horizontal equilibrium of the frame it follows that, for M_B acting alone, H_A also is zero. The bending moments in the members, would thus become

$$M_{AB} = -M_{BA} = M_{BC} = -M_B, \quad M_{CB} = M_{CD} = M_{DC} = 0.$$

On the other hand considering the effect of M_C alone and from the equilibrium of CD with $M_D = 0$, the reaction at D is given by $H_D = M_C/L_3$. From the horizontal equilibrium of the whole frame $H_A = M_C/L_3$. Considering the member AB and taking moments about B

$$M_A = \frac{M_c L_1}{L_3}$$

while $M_{BA} = M_{BC} = M_{DC} = 0$; $M_{CD} = -M_{CB} = -M_C$.
A similar procedure can be used for the effect of M_D acting

84

alone. Matrix B_r is thus given by:

$$\begin{bmatrix} M_{AB} \\ M_{BA} \\ M_{BC} \\ M_{CB} \\ M_{CD} \\ M_{DC} \end{bmatrix} = \begin{bmatrix} -1 & +L_1/L_3 & -L_1/L_3 \\ +1 & 0 & 0 \\ -1 & 0 & 0 \\ 0 & +1 & 0 \\ 0 & -1 & 0 \\ 0 & 0 & +1 \end{bmatrix} \begin{bmatrix} M_B \\ M_C \\ M_D \end{bmatrix} \qquad (2.75)$$

2.3.6. Automatic construction of the load transformation matrix

In the last section a number of examples were given for the construction of the load transformation matrix B. These are easy enough for the simple frames that were discussed. However for very large structures it becomes rather cumbersome, particularly when the matrix is required to be fed into a computer to carry out the matrix force analysis of a structure. It is advantageous, therefore, to devise an automatic procedure for the formation of the load transformation matrix that can be readily programmed for a computer.

In Section 2.2.2 reference was made to the duality of the various matrices used in the force and displacement methods. Use can be made of this dual nature and particularly, in the present instance, of the two equations (2.37) and (2.57), that is to say between $L = A' P$ and $P = B L$. For just as matrix B was partitioned into $[B_b \ B_r]$, matrix A' can also be partitioned as

$$L_b = A' P = [A'_b \ A'_r] \cdot \begin{bmatrix} P_b \\ P_r \end{bmatrix} \qquad (2.76)$$

where P_r are some of the member forces selected as the internal redundants. A'_r corresponds to P_r in that it has the same number of columns as P'_r, while A'_b is square. This implies that the partitioning is carried out in such a manner as to make A'_b have an equal number of rows and columns. Multiplying the right hand side of equations (2.76) gives

$$L_b = A'_b \ P_b + A'_r \ P_r$$

i.e. $\qquad A'_b \ P_b = L_b - A'_r \ P_r$

Hence: $\qquad P_b = (A_b')^{-1}.L_b - (A_b')^{-1}.A_r'.P_r$ $\qquad$ (2.77)

Also: $\qquad P_r = O.L_b + I\, P_r$

where O is a null matrix that has the same number of rows as there are redundants, while the number of its columns is the same as the order of L_b. Matrix I is the unit matrix with order r. Writing equations (2.77) in Matrix form

$$\begin{bmatrix} P_b \\ \cdots \\ P_r \end{bmatrix} = \begin{bmatrix} (A_b')^{-1} & -(A_b')^{-1}.A_r' \\ O & I \end{bmatrix} \begin{bmatrix} L_b \\ \cdots \\ P_r \end{bmatrix} \qquad (2.78)$$

It follows, by comparison of equations (2.78) and (2.65), that

$$B_b = \begin{bmatrix} (A_b')^{-1} \\ O \end{bmatrix} \quad \text{and} \quad B_r = \begin{bmatrix} (A_b')^{-1}.A_r' \\ I \end{bmatrix} \qquad (2.79)$$

The construction of $A' = [A_b'\ A_r']$ automatically follows easily from its derivations as given in Section 2.2.2. and utilising the scheme given in Section 2.2.4. for the presentation of joint conditions to the computer.

As an example the portal frame of Figure 2.14 is once again considered. Neglecting the axial stiffness of the members and noting that the sway of the two columns are equal and therefore one of them only should be considered, the joint displacement vector becomes $X = \{x_B\ \theta_B\ \theta_C\}$. The corresponding member displacements are $Z = \{v_{AB}\ \theta_{AB}\ \theta_{BA}\ \theta_{BC}\ \theta_{CB}\ v_{CD}\ \theta_{CD}\ \theta_{DC}\}$. Thus

$$\begin{bmatrix} v_{AB} \\ \theta_{AB} \\ \theta_{BA} \\ \theta_{BC} \\ \theta_{CB} \\ v_{CD} \\ \theta_{CD} \\ \theta_{DC} \end{bmatrix} = \begin{bmatrix} -1 & 0 & 0 \\ 0 & 0 & 0 \\ 0 & 1 & 0 \\ 0 & 1 & 0 \\ 0 & 0 & 1 \\ -1 & 0 & 0 \\ 0 & 0 & 1 \\ 0 & 0 & 0 \end{bmatrix} \begin{bmatrix} x_B \\ \theta_B \\ \theta_C \end{bmatrix},$$

$$
\begin{bmatrix} H_B \\ M_B \\ M_C \end{bmatrix} = \begin{bmatrix} -1 & 0 & 0 & 0 & 0 & -1 & 0 & 0 \\ 0 & 0 & 1 & 1 & 0 & 0 & 0 & 0 \\ 0 & 0 & 0 & 0 & 1 & 0 & 1 & 0 \end{bmatrix} \begin{bmatrix} S_{AB} \\ M_{AB} \\ M_{BA} \\ M_{BC} \\ M_{CB} \\ S_{CD} \\ M_{CD} \\ M_{DC} \end{bmatrix}
$$

In the load transformation process $P = B.L$ the shear forces S_{AB} and S_{CD} are not required. These can, therefore, be excluded. From the equilibrium of AB and CD, it is found that

$$
S_{AB} = -\frac{M_{AB}}{L_1} - \frac{M_{BA}}{L_1}, \quad S_{CD} = \frac{-M_{CD}}{L_3} - \frac{M_{DC}}{L_3}
$$

Thus substituting for $-S_{AB}$ and $-S_{CD}$ in $L = A'.P$

$$
\begin{bmatrix} H_B \\ M_B \\ M_C \end{bmatrix} = \begin{bmatrix} 1/L_1 & 1/L_1 & 0 & 0 & 1/L_3 & 1/L_3 \\ 0 & 1 & 1 & 0 & 0 & 0 \\ 0 & 0 & 0 & 1 & 1 & 0 \end{bmatrix} \begin{bmatrix} M_{AB} \\ M_{BA} \\ M_{BC} \\ M_{CB} \\ M_{CD} \\ M_{DC} \end{bmatrix} \quad (2.80)
$$

We notice in equation (2.80), that if the first three columns are partitioned from the rest, the resulting 3×3 matrix is singular. This indicates that the last three member forces M_{CB}, M_{CD} and M_{DC} cannot be selected as the redundants because M_{CB} and M_{CD} amount to only one redundant. A re-arrangement of the columns is therefore, required. For instance M_{BA}, M_{CB} and M_{DC} can be selected resulting in

$$
\begin{bmatrix} H_A \\ M_B \\ M_C \end{bmatrix} = \left[\begin{array}{ccc:ccc} 1/L_1 & 0 & 1/L_3 & 1/L_1 & 0 & 1/L_3 \\ 0 & 1 & 0 & 1 & 0 & 0 \\ 0 & 0 & 1 & 0 & 1 & 0 \end{array} \right] \begin{bmatrix} M_{AB} \\ M_{BC} \\ M_{CD} \\ \hdashline M_{BA} \\ M_{CB} \\ M_{DC} \end{bmatrix}
$$

87

and

$$A_b' = \begin{bmatrix} 1/L_1 & 0 & 1/L_3 \\ 0 & 1 & 0 \\ 0 & 0 & 1 \end{bmatrix}, \qquad A_r' = \begin{bmatrix} 1/L_1 & 0 & 1/L_3 \\ 1 & 0 & 0 \\ 0 & 1 & 0 \end{bmatrix}$$

Inverting A_b' gives: $(A_b')^{-1} = \begin{bmatrix} L_1 & 0 & -L_1/L_3 \\ 0 & 1 & 0 \\ 0 & 0 & 1 \end{bmatrix}$

and thus

$$(-1)(A_b')^{-1}.A_r' = \begin{bmatrix} -L_1 & 0 & L_1/L_3 \\ 0 & -1 & 0 \\ 0 & 0 & -1 \end{bmatrix} \begin{bmatrix} 1/L_1 & 0 & 1/L_3 \\ 1 & 0 & 0 \\ 0 & 1 & 0 \end{bmatrix} =$$

$$= \begin{bmatrix} -1 & L_1/L_3 & -L_1/L_3 \\ -1 & 0 & 0 \\ 0 & -1 & 0 \end{bmatrix}$$

substituting in equations (2.78) gives

$$\begin{bmatrix} M_{AB} \\ M_{BC} \\ M_{CD} \\ \cdots \\ M_{BA} \\ M_{CB} \\ M_{DC} \end{bmatrix} = \begin{bmatrix} L_1 & 0 & -L_1/L_3 & -1 & L_1/L_3 & -L_1/L_3 \\ 0 & 1 & 0 & -1 & 0 & 0 \\ 0 & 0 & 1 & 0 & -1 & 0 \\ 0 & 0 & 0 & 1 & 0 & 0 \\ 0 & 0 & 0 & 0 & 1 & 0 \\ 0 & 0 & 0 & 0 & 0 & 1 \end{bmatrix} \begin{bmatrix} H_B \\ M_B \\ M_C \\ \cdots \\ M_{BA} \\ M_{CB} \\ M_{DC} \end{bmatrix}$$

Reorganising these equations so that the left hand side is the same as the member force vector given originally by equations (2.80) results in

$$\begin{bmatrix} M_{AB} \\ \\ M_{BA} \\ M_{BC} \\ M_{CB} \\ M_{CD} \\ M_{DC} \end{bmatrix} = \begin{bmatrix} L_1 & 0 & -L_1/L_3 & -1 & L_1/L_3 & -L_1/L_3 \\ 0 & 0 & 0 & 1 & 0 & 0 \\ 0 & 1 & 0 & -1 & 0 & 0 \\ 0 & 0 & 0 & 0 & 1 & 0 \\ 0 & 0 & 1 & 0 & -1 & 0 \\ 0 & 0 & 0 & 0 & 0 & 1 \end{bmatrix} \begin{bmatrix} H_B \\ M_B \\ M_C \\ \cdots \\ M_{BA} \\ M_{CB} \\ M_{DC} \end{bmatrix}$$

88

Hence with H_B, M_B and M_C acting on the frame

$$B_b = \begin{bmatrix} L_1 & 0 & -L_1/L_3 \\ 0 & 0 & 0 \\ 0 & 1 & 0 \\ 0 & 0 & 0 \\ 0 & 0 & 1 \\ 0 & 0 & 0 \end{bmatrix} \quad \text{and} \quad B_r = \begin{bmatrix} -1 & L_1/L_3 & -L_1/L_3 \\ 1 & 0 & 0 \\ -1 & 0 & 0 \\ 0 & 1 & 0 \\ 0 & -1 & 0 \\ 0 & 0 & 1 \end{bmatrix}$$

The matrix B_r thus obtained is the same as that given for the same portal at the end of the last section.

EXERCISES

1. Using matrix displacement method, calculate the bending moments M_{AB}, M_{BA}, M_{BC}, and M_{CB} for the column of example 1.1, shown in Figure 1.11.

2. The co-ordinates (X, Y, Z) of the pin jointed space frame of Figure 2.15 are given in brackets in the figure. The arrows for

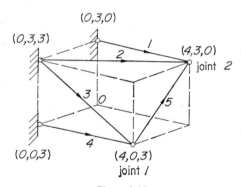

Figure 2.15

each member point to the second end of the member and the figure next to each arrow is the member number. Construct the displacement transformation matrix for the structure.

Answer.

$$
\begin{bmatrix}
0 & 0 & 0 & 1 & 0 & 0 \\
0 & 0 & 0 & 0.8 & 0 & -0.6 \\
0.8 & -0.6 & 0 & 0 & 0 & 0 \\
1 & 0 & 0 & 0 & 0 & 0 \\
0 & -1/\sqrt{2} & 1/\sqrt{2} & 0 & 1/\sqrt{2} & -1/\sqrt{2}
\end{bmatrix}
$$

3. Repeat the problem for the pin-jointed space frame of Figure 2.16.

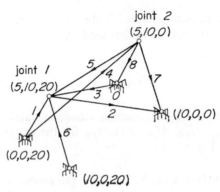

Figure 2.16

Answer.

$$
\begin{bmatrix}
0.447 & 0.895 & 0 & 0 & 0 & 0 \\
-0.213 & 0.425 & 0.850 & 0 & 0 & 0 \\
0.213 & 0.425 & 0.850 & 0 & 0 & 0 \\
0 & 0 & 0 & 0.213 & 0.425 & 0.850 \\
0 & 0 & 1 & 0 & 0 & -1 \\
-0.447 & 0.895 & 0 & 0 & 0 & 0 \\
0 & 0 & 0 & -0.447 & 0.895 & 0 \\
0 & 0 & 0 & 0.447 & 0.895 & 0
\end{bmatrix}
$$

4. The beam *ABC* of Figure 2.17 is built in at *A* and pinned at *C*. At point *B* the beam is connected to the support at *D* by a vertical spring of axial stiffness *k*. Construct the overall stiffness matrix for the system. Calculate the vertical displacement of point *B* when a vertical load *V* is applied there.

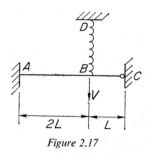

Figure 2.17

Answer.

$$\begin{bmatrix} e_1+e_2 & f_2 & d_1-d_2 \\ f_2 & e_2 & -d_2 \\ d_1-d_2 & -d_2 & b_1+b_2+k \end{bmatrix}, \qquad X. = \{\theta_B\ \theta_C\ y_B\}$$

$$y_B = V/[t(d_1+d_2+f_2/e_2)+b_1+b_2+k-d_2^2/e_2),]\quad \text{where}$$

$$t = [d_1-d_2+d_2f_2/e_2]/[-e_1-e_2+f_2^2/e_2]$$

5. The symmetrical pitched roof frame of Figure 2.18 is subject to two moments M at B and D. Sketch the deformed shape of the frame and indicate the joint deformations that are significant.

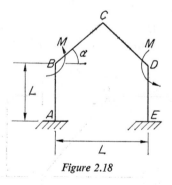

Figure 2.18

Construct the overall stiffness matrix for this mode of deformation making the order of this matrix as small as possible. If the horizontal sway at B is $0\cdot1L$, calculate the value of M. EI is constant and the axial stiffness EA/L of the members can be neglected.

Answer.
$$\theta_C = -0\cdot5\theta_B, \qquad X = \{x_B\ \theta_B\}$$
$$K = \begin{bmatrix} b_1 & -d_1 \\ -d_1 & e_1 + 3e_2/4 \end{bmatrix}$$
$$M = -0\cdot2EI(1+6\cos\alpha)/L$$

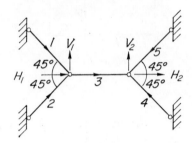

Figure 2.19

6. All the members of the pin-jointed plane frame in Figure 2.19 have the same lengths and cross-sectional properties. The member f orces are

$$\{P_1\ P_2\ P_3\ P_4\ P_5\} = \frac{EA}{\sqrt{2L}}\{2\ \ 3\ \ -3\sqrt2\ \ 3\ \ -1\}$$

Calculate the external forces at joints 1 and 2.

Answer. $\qquad \{H_1\ V_1\ H_2\ V_2\} = \dfrac{EA}{L}\{3\ 1\ 0\ -2\}$

7. The bent bar *ABCD* is fixed at *A* and attached at *D* to a spring of rotational stiffness *k* as shown in Figure 2.20. A moment

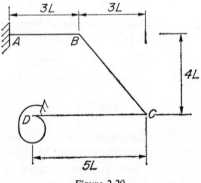

Figure 2.20

92

M is applied at D so that the spring rotation there is 0·1 rad. Calculate the vertical deflection at C. EI is constant. Use matrix force method. (Hint $M = 0·1k$)

Answer. $$v = -2·1kL^2/EI$$

8. Using matrix force method calculate the vertical displacement at point B of the pin-jointed frame in Figure 2.21, due to a unit vertical load acting at A. All the members are made out of the same section.

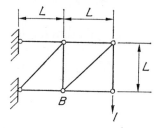

Figure 2.21

Answer. $$y_B = (1+4\sqrt{2})L/EA$$

9. Using matrix force method, derive expressions for the horizontal and vertical displacements at point D in the structure of Figure 2.22.

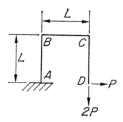

Figure 2.22

Answer. $$y = 5PL^3/3EI, \quad x = -PL^3/3EI$$

10. Repeat example 1.1 by using matrix force method.

11. Prove that the forces P developed in the members of a statically indeterminate structure, due to a set of lack of fit X_r in the redundant members, are given by

$$P = B_r(B_r' \, f \, B_r)^{-1}X_r$$

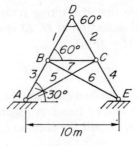

Figure 2.23

The member *BC* of the pin-jointed frame in Figure 2.23 is 10 mm too short. Calculate the forces developed in the members of the frame when *BC* is forced into its position. Take $E = 207$ kN/mm² and the cross-sectional area of each member as 100 mm².

Answer.

$$\left\{0 \quad 0 \quad \frac{207}{41} \quad \frac{207}{41} \quad \frac{-207\sqrt{3}}{41} \quad \frac{-207\sqrt{3}}{41} \quad \frac{207}{20 \cdot 5}\right\}$$

CHAPTER 3

The Stability Functions

3.1. THE EFFECT OF AXIAL LOAD ON THE STIFFNESS OF A MEMBER

In the previous chapter, where the linear analysis of structures was considered, it was assumed that both the stiffness and the flexibility of the members, as well as those of the structure, are constants. This assumption is valid provided that there are no axial forces in the members or that these remain at their constant values during the process of analysing a structure. Once the axial forces in the members of a structure vary, their individual stiffnesses change and so does the overall stiffness of the structure. In Chapter 1 it was shown that the influence of the axial forces in the member is a significant cause of non-linearity in the behaviour of structures. It is therefore necessary to study the properties of the various members of a structure with varying stiffness due to the presence of axial forces. In this chapter this will be done with reference to a single member.

3.1.1. Rotation of a member

Consider first a propped cantilever AB in Figure 3.1(a) subject, at end A, to an axial force p, a bending moment M_{AB} and a shear force S. In the absence of the axial force, it is already known that

$$M_{AB} = 4\frac{EI}{L}\theta_A = 4k\theta_A,$$

$$M_{BA} = 2k\theta_A \qquad (3.1)$$

The quantity $M_{AB}/\theta_A = 4k$ is the moment required at end A to cause a unit rotation there and is called the stiffness of the member against rotation of A. The moment carried over to the second end of the member is half the applied moment.

For values of p other than zero the bending moments would be different from those given in equations (3.1). However these moments can be expressed in a manner analogous to (3.1) as

$$M_{AB} = sk\theta_A,$$

$$M_{BA} = csk\theta_A = cM_{AB} \qquad (3.2)$$

where s is known as the stiffness factor and c is the carry-over factor. Expressions for s and c are derived in Section 3.2, but it is useful to continue with the analysis of the member as this clarifies the meaning of the stiffness of a member.

Taking moments about end B of the member gives

$$SL + M_{AB} + M_{BA} = 0$$

i.e. $$S = \frac{-M_{AB} - M_{BA}}{L} = \frac{-sk\theta_A - csk\theta_A}{L} = \frac{-s(1+c)k\theta_A}{L} \qquad (3.3)$$

Similarly if end B of the member is pinned while end A is fixed as shown in Figure 3.1(b)

$$M_{BA} = sk\theta_B,$$

$$M_{AB} = csk\theta_B \qquad (3.4)$$

The condition of pure rotation, without sway, of a member is obtained by direct addition of the above two cases, as shown in Figure 3.1(c). Thus from equations (3.2) and (3.4):

$$M_{AB} = sk\theta_A + csk\theta_B$$

$$M_{BA} = csk\theta_A + sk\theta_B \qquad (3.5)$$

while $$S = -s(1+c)k(\theta_A + \theta_B)/L$$

96

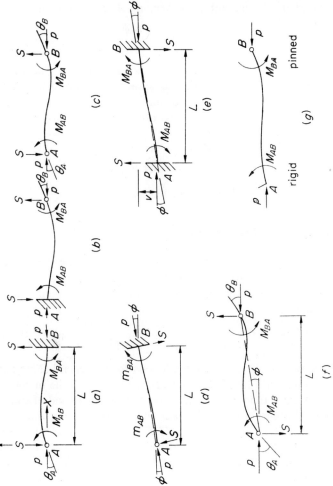

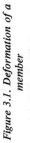

Figure 3.1. Deformation of a member

(a) Rotation of A +
(b) Rotation of B =
(c) Pure rotation
(d) Rotation of A for an inclined member
(e) Pure sway
(f) A general state of sway
(g) General sway of a member with one end pinned

97

3.1.2. Sway of a member

Consider next the case of an inclined member and apply a moment m_{AB}, as shown in Figure 3.1(d), so that the tangent at A to the member becomes horizontal and the rotation at that end is $-\phi$. Then using equations (3.2):

$$m_{AB} = -sk\phi$$
$$m_{BA} = -csk\phi$$

The application of another moment m'_{BA} at B in order to make that end also horizontal, while keeping A fixed results in

$$m'_{BA} = -sk\phi$$
$$m'_{AB} = -csk\phi$$

The member now has both ends horizontal with one of them subsided relative to the other as shown in Figure 3.1(e). This is the state of pure sway without any rotation. The final moments at the ends of the member are obtained by adding m and m'. Thus

$$M_{AB} = -s(1+c)k\phi$$
$$M_{BA} = -s(1+c)k\phi$$

$$(3.6)$$

where $\phi = v/L$ as shown in the figure.

3.1.3. A state of general sway

This state of deformation is the result of adding the two cases of pure rotation and pure sway for a member. This is shown in Figure 3.1(f) and the end moments are obtained from equations (3.5) and (3.6) as

$$M_{AB} = sk\theta_A + csk\theta_B - s(1+c)k\phi$$
$$M_{BA} = csk\theta_A + sk\theta_B - s(1+c)k\phi$$

$$(3.7)$$

These two equations are analogous to the general slope deflection equations (2.5). The moments at the ends of the member are in both cases functions of the sway of the member and of end rotations. However, due to the presence of axial loads, the bending moments given by equations (3.7) are different from those of equations (2.5). It is readily seen that this difference is two-fold. Firstly, there is the influence of the axial loads on the member curvature expressed in terms of the end rotations as, for instance

98

for M_{AB}, $sk\theta_A + csk\theta_B$. Secondly, there is the effect of the axial loads on the joint translation expressed as $s(1+c)k\phi$ for the member.

3.1.4. General sway of a member with hinges

A more general case is that of a member of a frame having a rigid joint at one end and a pin at the other. This is shown in Figure 3.1(g). In this case, because of the pin at end B, the bending moment remains zero at that end and the second of equations (3.7) becomes

$$M_{BA} = csk\theta_A + sk\theta_B - s(1+c)k\phi = 0$$

This gives

$$csk\theta_B = cs(1+c)k\phi - c^2 sk\theta_A,\qquad(3.8)$$

and substituting for $csk\theta_B$ in the first of equations (3.7) results in

$$M_{AB} = s(1-c^2)k\theta_A - s(1-c^2)k\phi\qquad(3.9)$$

For the case when the member has no lateral sway, the second term of equation (3.9) vanishes, giving

$$M_{AB} = s(1-c^2)k\theta_A,\qquad(3.10)$$

and equation (3.8) gives the value of θ_B as:

$$\theta_B = -c\theta_A,\qquad(3.11)$$

where, as stated, c is the carry-over factor.

3.2. THE STABILITY FUNCTIONS USED IN THE MATRIX DISPLACEMENT METHOD

In the propped cantilever of Figure 3.1(a) the member forces at A are the axial load p, the shear force S and the bending moment M_{AB}. Those at the other end are, respectively, p, S and the induced moment M_{BA}. Point A is chosen as the origin for the axes X and Y of the member, adopting the usual sign conventions as shown in the figure. The bending moment at any other point, of distance

x from A, is obtained by taking moments about that point, as

$$M_x = -EI\frac{d^2y}{dx^2} = (M_{AB}+Sx+py) \qquad (3.12)$$

where EI is the flexural rigidity of the member and y is the lateral displacement of the point. It is advantageous to express the axial load p as a factor ϱ of the Euler critical load $P_E = \pi^2EI/L^2$ for a pin-ended member of the same length and stiffness as those of AB. That is

$$p = \varrho P_E = \frac{\varrho\pi^2EI}{L^2} = \frac{\varrho\pi^2k}{L} \qquad (3.13)$$

or

$$\varrho = \frac{p}{P_E}$$

The values of the bending moment M_{AB} and the shear force S are given in equations (3.2) and (3.3) respectively. Substituting for these and for the axial force p from (3.13) into equation (3.12) gives the value of the curvature of the member as

$$\frac{d^2y}{dx^2} = -\frac{1}{EI}\left[sk\theta_A - \frac{s(1+c)kx}{L}\theta_A + \frac{\varrho\pi^2k}{L}y\right] \qquad (3.14)$$

The solution of this equation is quoted as

$$y = A \sin \pi\sqrt{\varrho}\,\frac{x}{L} + B \cos \pi\sqrt{\varrho}\,\frac{x}{L} + \frac{L}{\pi^2\varrho k}\left[\frac{s(1+c)k\theta_A x}{L} - sk\theta_A\right] \qquad (3.15)$$

which can be checked by differentiating y twice with respect to x, thus:

$$\frac{dy}{dx} = \frac{A\pi\sqrt{\varrho}}{L}\cos\frac{\pi\sqrt{\varrho}}{L}x - \frac{B\pi\sqrt{\varrho}}{L}\sin\frac{\pi\sqrt{\varrho}}{L}x + \frac{L}{\pi^2\varrho k}\left[\frac{s(1+c)k\theta_A}{L}\right] \qquad (3.16)$$

and

$$\frac{d^2y}{dx^2} = -\frac{\pi^2\varrho}{L^2}\left(A \sin\frac{\pi\sqrt{\varrho}}{L}x + B \cos\frac{\pi\sqrt{\varrho}}{L}x\right), \qquad (3.17)$$

but since from equation (3.15)

$$A \sin\frac{\pi\sqrt{\varrho}}{L}x + B \cos\frac{\pi\sqrt{\varrho}}{L}x = y - \frac{L}{\pi^2\varrho k}\left[\frac{s(1+c)k\theta_A x}{L} - sk\theta_A\right],$$

100

equations (3.14) follow directly. The constants A and B are found from the boundary conditions, that are at $x = 0$ or $x = L$, $y = 0$ giving

$$A = -\frac{M_{AB} \cot 2\alpha + M_{BA} \operatorname{cosec} 2\alpha}{4\alpha^2 k/L},$$

$$B = \frac{M_{AB}L}{4\alpha^2 k}$$

where
$$\alpha = 0 \cdot 5\pi\sqrt{\varrho} \qquad (3.18)$$

Further at $x = L$, the slope $dy/dx = 0$, therefore, substituting for A and B in equation (3.16) with the value of $x = L$ and using (3.18) it follows that

$$c = \frac{2\alpha - \sin 2\alpha}{\sin 2\alpha - 2\alpha \cos 2\alpha} \qquad (3.19)$$

Further, since when $x = 0$, $dy/dx = \theta_A$ and using (3.16) again together with (3.19) we obtain

$$s = \frac{(1 - 2\alpha \cot 2\alpha)\alpha}{\tan \alpha - \alpha} \qquad (3.20)$$

Defining a function $\phi_1 = (\pi\sqrt{\varrho})/2 \cot (\pi\sqrt{\varrho})/2 = \alpha \cot \alpha$ and substituting this in (3.20) and (3.19) they become

$$s = \frac{0 \cdot 25\pi^2\varrho + \phi_1 - \phi_1^2}{1 - \phi_1}$$

$$c = \frac{1}{4s} \cdot \frac{\pi^2\varrho - 4\phi_1 + 4\phi_1^2}{1 - \phi_1} \qquad (3.21)$$

In Figure 3.2 the values of s and c are plotted against ϱ where it is noticed that for negative values of ϱ; s is more than 4. That is to say as the tensile force in a member increases the member becomes stiffer. As ϱ increases, the value of s decreases until at $\varrho = 0$, $s = 4$. At $\varrho \simeq 2 \cdot 046$, $s = 0$. This means that the member has zero stiffness and a small disturbing force causes elastic buckling of the member. On the other hand c increases with ϱ until at $\varrho = 0$, $c = 0 \cdot 5$. The value of c tends to infinity as $\varrho \to 2 \cdot 046$. After this value of ϱ, c becomes negative. s and c are tabulated in Appendix 1.

Let us now consider the case of pure sway as shown in Figure 3.1(e). The shearing force S at the end of the member is obtained by taking moments about one end B giving

$$SL = -M_{AB} - M_{BA} - pv$$

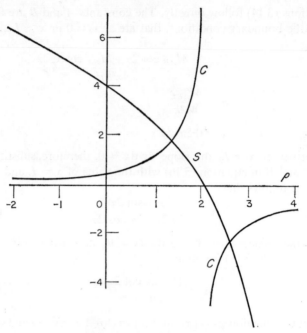

Figure 3.2. S and C Functions

Substituting in this equation the values of M_{AB} and M_{BA} from equations (3.6), with $v = \phi L$ and using equations (3.13) for $p = \varrho \pi^2 k / L$ and (3.21) gives

$$S = \frac{2s(1+c)\phi_1 kv}{L^2} \tag{3.22}$$

Finally, substituting for $s(1+c)kv/L = 0 \cdot 5SL/\phi_1$ from equation (3.22) into (3.6) results in

$$M_{AB} = M_{BA} = -\frac{0 \cdot 5SL}{\phi_1} \tag{3.23}$$

or

$$S = -\frac{2M_{AB}\phi_1}{L}$$

With the presence of an axial force, these equations show the relationship between the shear force in a member in pure sway and the end moments. They also show that these values can be calculated using the stability function ϕ_1 alone.

It is advantageous, in the construction of the stiffness matrix

102

of a member, to introduce four more stability functions all of which can be expressed in terms of α. For simplicity, however, they are given here in terms of α and each other. These are

$$\left.\begin{aligned}
\phi_2 &= \frac{\alpha^2}{3(1-\phi_1)} \\[2mm]
\phi_3 &= \frac{3\phi_2+\phi_1}{4} \\[2mm]
\phi_4 &= \frac{3\phi_2-\phi_1}{2}
\end{aligned}\right\} \qquad (3.24)$$

and
$$\phi_5 = \phi_1\phi_2$$

These ϕ functions are due to R. K. Livesley[2] and can be used to derive the slope-deflection equations of general members. In Figure 3.3 these functions are plotted against ϱ and are given in

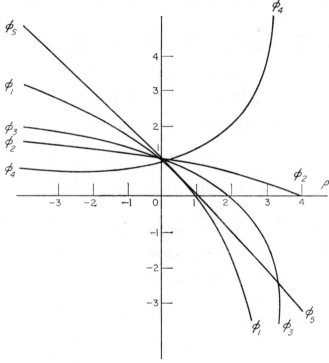

Figure 3.3. Livesley's functions

tabular form in Appendix 1. It is a useful common feature of the ϕ functions that in the absence of any axial load in a member, when $\varrho = 0$, these functions all take the value of unity.

Using these functions, the slope deflection equations (3.5), for the case of pure rotation, become

$$M_{AB} = 4\phi_3 k\theta_A + 2\phi_4 k\theta_B,$$
$$M_{BA} = 2\phi_4 k\theta_A + 4\phi_3 k\theta_B, \qquad (3.25)$$
$$S = -6\phi_2 k(\theta_A + \theta_B)/L$$

The reader is now in a position to develop expressions for the member forces for the other modes of deformation discussed in section (3.1).

3.3. THE STABILITY FUNCTIONS USED IN MATRIX FORCE METHOD

The end moments of a member in pure rotation are given by the first two of equations (3.25). Writing these in matrix form

$$\begin{bmatrix} M_{AB} \\ M_{BA} \end{bmatrix} = k \begin{bmatrix} 4\phi_3 & 2\phi_4 \\ 2\phi_4 & 4\phi_3 \end{bmatrix} \begin{bmatrix} \theta_A \\ \theta_B \end{bmatrix}$$

Inverting the stiffness matrix of these equations

$$k^{-1}\begin{bmatrix} 4\phi_3 & 2\phi_4 \\ 2\phi_4 & 4\phi_3 \end{bmatrix}^{-1} = \frac{1}{k}\begin{bmatrix} \phi_6/3 & -\phi_7/6 \\ -\phi_7/6 & \phi_6/3 \end{bmatrix},$$

where
$$\phi_6 = \frac{\phi_3}{\phi_2(2\phi_3 - \phi_4)}, \qquad (3.26)$$

and
$$\phi_7 = \phi_4\phi_6/\phi_3$$

Thus the rotations θ_A and θ_B are given by

$$\begin{bmatrix} \theta_A \\ \theta_B \end{bmatrix} = \frac{L}{6EI}\begin{bmatrix} 2\phi_6 & -\phi_7 \\ -\phi_7 & 2\phi_6 \end{bmatrix}\begin{bmatrix} M_{AB} \\ M_{BA} \end{bmatrix} \qquad (3.27)$$

The functions ϕ_6 and ϕ_7 were developed by Berry[3] in 1916. These

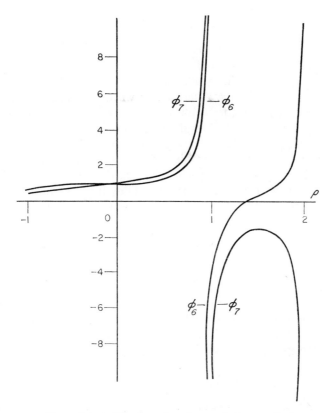

Figure 3.4. Stability Functions for matrix force method

were the first stability functions employed in the non-linear analysis of structures by a force method.

Graphs of ϕ_6 and ϕ_7 against ϱ are shown in Figure 3.4 where it can be seen that they also become unity in the absence of any axial load, thus converting equations (3.27) into (2.52)

3.4. RELATIONSHIPS BETWEEN THE STABILITY FUNCTIONS

It is apparent that the various stability functions are closely related to each other. In many cases, however, the reader may prefer one particular set of these functions. Indeed certain problems

105

become clearer with one set of stability functions as compared to another. For instance, using the stiffness approach manually without the use of matrix procedures, it is more convenient to use the s and c functions, particularly since the user realises during the various operations that these functions are the familiar stiffness and carry over factors. Alternatively when a computer is being employed in conjunction with a matrix method, the ϕ functions become more efficient. For these reasons it is advantageous to prepare a list that relates the various functions to each other.

Before doing this, the meaning of a few more stability functions is explained so that they may also be added to the list for completeness. One such function is the stability function m. This is in fact the reciprocal of ϕ_1 used in problems associated with the sway of a member. Thus

$$m = \frac{1}{\phi_1} \qquad (3.28)$$

Two further stability functions that are found of some use are the 'no shear functions', n and o. These are obtained by superimposing a pure sway on the rotation of one end of a member as shown in Figure 3.5. The final bending moments at the ends of the members are obtained by adding the separate values as given

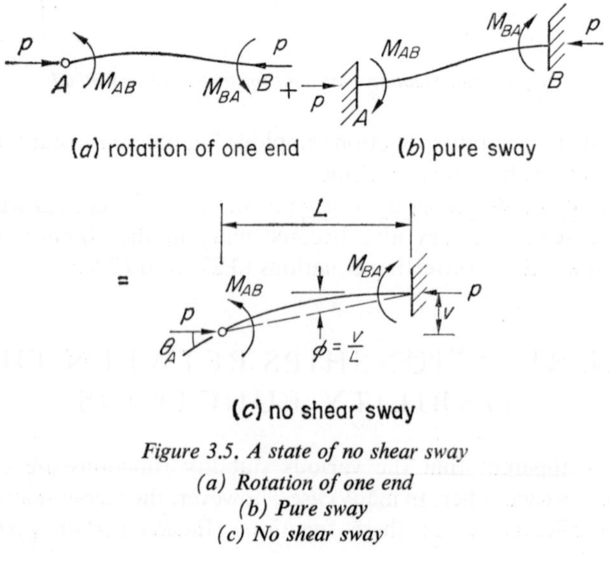

(a) rotation of one end   (b) pure sway

(c) no shear sway

Figure 3.5. A state of no shear sway
(a) Rotation of one end
(b) Pure sway
(c) No shear sway

in equations (3.2) and (3.6), i.e.

$$M_{AB} = sk\theta_A - s(1+c)k\theta = nk\phi_A$$
$$M_{BA} = csk\theta_A - s(1+c)k\phi = -ok\phi_A$$
(3.29)

The shear force S in the member is obtained from equations (3.3) and (3.22) using (3.28) as

$$S = -\frac{s(1+c)k\theta_A}{L} + \frac{2s(1+c)k\phi}{mL} = 0$$

This is true provided that ϕ is $0.5m\theta_A$ and hence

$$n = s\left[1 - \frac{m(1+c)}{2}\right]$$
$$o = s\left[\frac{m(1+c)}{2} - c\right]$$
(3.30)

The relationships between the various functions are straightforward and it is worthwhile for the reader to derive them. These are

$$\varrho = p/P_E = \frac{PL^2}{\pi^2 EI}$$

$$p = \varrho P_E = \frac{\varrho\pi^2 k}{L}$$

$$\phi_1 = \frac{\pi}{2}\sqrt{\varrho}\cot\frac{\pi\sqrt{\varrho}}{2} = \alpha\cot\alpha = \frac{1}{m}$$

$$\phi_2 = \frac{\alpha^2}{3(1-\phi_1)} = \frac{s(1+c)}{6}$$

$$\phi_3 = \frac{3\phi_2 + \phi_1}{4} = s/4$$

$$\phi_4 = \frac{3\phi_2 - \phi_1}{2} = \frac{1}{2}sc$$

$$\phi_5 = \phi_1\phi_2 = \frac{s(1+c)}{6m}$$

$$\phi_6 = \frac{\phi_3}{\phi_2(2\phi_3 - \phi_4)} = \frac{3}{s(1-c^2)}$$

$$\phi_7 = \phi_4\phi_6/\phi_3 = \frac{6c}{s(1-c^2)}$$

107

$$s = \frac{(1-2\alpha\cot 2\alpha)\alpha}{\tan\alpha - \alpha} = \frac{0\cdot 25\pi^2\varrho + \phi_1 - \phi_1^2}{1-\phi_1} = 4\phi_3 = \frac{12\phi_6}{4\phi_6^2 - \phi_7^2}$$

$$c = \frac{2\alpha - \sin 2\alpha}{\sin 2\alpha - 2\alpha\cos 2\alpha} = \frac{1}{4s}\left(\frac{\pi^2\varrho - 4\phi_1 + 4\phi_1^2}{1-\phi_1}\right) = \frac{1}{2}\phi_4/\phi_3$$

$$= \frac{\phi_7}{2\phi_6}$$

$$m = \frac{2s(1+c)}{2s(1+c) - \pi^2\varrho} = \frac{1}{\phi_1}$$

$$n = s\left[1 - \frac{m(1+c)}{2}\right] = 4\phi_3 - 3\phi_2/\phi_1$$

$$o = s\left[\frac{m(1+c)}{2} - c\right] = \frac{3\phi_2}{\phi_1} - 2\phi_4$$

$$s(1-c^2) = \frac{\pi^2\varrho}{1-n}$$

$$s(1-c) = 2/m = 2\phi_1$$

$$s(1+c) = \frac{o-n}{m-1} = \frac{0\cdot 5m\pi^2\varrho}{m-1} = 6\phi_2$$

$$s = \frac{1-n}{m-1}$$

$$1 - mn = m\pi^2\varrho/4$$

$$o + n = \frac{2}{m}$$

$$o - n = \frac{m\pi^2\varrho}{2}$$

$$o^2 - n^2 = \pi^2\varrho$$

As a practical application of this list the column AB of Figure 3.6 will be analysed. This is a typical ground floor column of multi-storey frame which is pinned to the foundation at A as shown in the figure. The column is subject to an axial load p, a shear force S and an applied moment M_{BA} at end B. It is required to calculate the end rotations θ_B and θ_A in terms of these forces only.

The bending moment at A due to the end rotation θ_A and a pure sway is given by equation (3.29) as $nk\theta_A$. On the other hand a

108

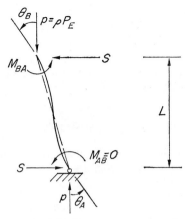

Figure 3.6. The forces acting on the column

bending moment of $-ok\theta_B$ is induced at A which is caused by a rotation θ_B at B. Finally, the application of the shear force S, causes at end A, a bending moment of $-0\cdot5SL/\phi_1 = -SLm/2$. The total moment in the member thus becomes

$$M_{AB} = nk\theta_A - ok\theta_B - \frac{SLm}{2}$$

Similarly   (3.31)

$$M_{BA} = -ok\theta_A + nk\theta_B - \frac{SLm}{2}$$

Equations (3.31) are in fact the same as equations (3.7) and express the state of general deformation of a member using n, o and m functions.

Because there is a hinge at A, the bending moment there is zero and the first of equations (3.31) gives

$$k\theta_A = \frac{ok\theta_B}{n} + \frac{SLm}{2n},$$ (3.32)

substituting this in the second of equations (3.31), we obtain

$$M_{BA} = \frac{n^2 - o^2}{n} k\theta_B - \frac{mSL}{2}\left(\frac{n+o}{n}\right).$$

Now from the list of stability functions it is noticed that $n^2 - o^2 =$

109

$= \pi_{\varrho}^2$ and $o+n = 2/m$. It follows that

$$M_{BA} = -\frac{\pi^2 \varrho k \theta_B + SL}{n}$$

and since $n = 4\phi_3 - 3\phi_2/\phi_1$, the bending moment at B in terms of the ϕ functions becomes

$$M_{BA} = -\frac{\phi_1(\pi^2 \varrho k \theta_B + SL)}{4\phi_3\phi_1 - 3\phi_2}$$

Once again using the list to eliminate the ϕ functions and to express M_{BA} in terms of θ, ϱ, S and α

$$M_{BA} = -\frac{\cot \alpha (\pi^2 \varrho k \theta_B + SL)}{\alpha(-1 + \cot^2 \alpha)}$$

Hence,
$$\theta_B = -\frac{\pi\sqrt{\varrho} M_{BA}\cos \pi\sqrt{\varrho} + SL \sin \pi\sqrt{\varrho}}{\pi^2 \varrho k \sin \pi\sqrt{\varrho}}$$

which is the required expression for θ_B. The reader may obtain a similar expression for θ_A' using equation (3.32).

3.5 GENERAL DEFORMATION OF A MEMBER WITH PLASTIC HINGES

During the proportional loading of a structure, the situation arises when a plastic hinge develops at one or both ends of a deformed member. The rotation of an end of a member after the development of a plastic hinge is partly due to the elastic deformation of that end and partly due to the hinge rotation. The treatment of these rotations separately is given in Chapter 6 when the elastic-plastic analysis of structures is presented. However, it is convenient to make use of the fact that the bending moment at a plastic hinge is constant and equal to the fully plastic moment of the section M_p, and eliminate the unknown end rotation there.

The method used in the example of the column in section 3.4 can be employed here using equations (3.31). In order to familiarise the reader with the various stability functions derived earlier, the general equations (3.7) are selected for this purpose. Thus, for a member in a state of general sway

$$M_{AB} = sk\theta_A + csk\theta_B - s(1+c)k\phi,$$
$$M_{BA} = csk\theta_A + sk\theta_B - s(1+c)k\phi$$

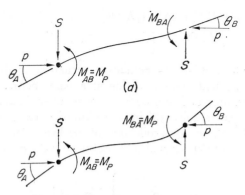

Figure 3.7. A member with plastic hinges
(a) A plastic hinge at end A
(b) Plastic hinges at both ends

and the shear force in the member is

$$S = -s(1+c)\frac{k}{L}\theta_B - s(1+c)\frac{k}{L}\theta_B + \frac{2s(1+c)k}{mL}\phi$$

If a hinge exists at end A of the member as shown in Figure 3.7(a), $M_{AB} = M_p$. Hence

$$sk\theta_A + csk\theta_B - s(1+c)k\phi = M_p$$

i.e.

$$csk\theta_A = cM_p - c^2sk\theta_B + sc(1+c)k\phi$$

and the bending moment at the second end becomes

$$M_{BA} = cM_p + s(1-c^2)k(\theta_B - \phi)$$

or

$$\theta_B - \phi = \frac{1}{3k}(\phi_6 M_{BA} - 0 \cdot 5\phi_7 M_p) \tag{3.33}$$

The shear force in the member is

$$S = \frac{s(1+c)}{L}\left[k\phi\left(\frac{2}{m} - 1 - c\right) - k\theta_B(1-c) - M_p/s\right] \tag{3.34}$$

If a second hinge is introduced at end B of the member as shown in Figure 3.7(b), then $M_{BA} = M_{AB} = M_p$. The shear force in the member is then obtained by taking moments about one end thus

$$SL = -2M_p - pL\phi \tag{3.35}$$

111

3.6. EFFECT OF LATERAL LOADING AND GUSSET PLATES

An advantage of matrix methods of structural analysis lies in the fact that lateral loading acting at a point between the two ends of a member does not complicate the analysis qualitatively. Such a load only requires the introduction of an extra joint to the structure at the point of application of the load. This method of introducing extra joints can also be useful when the forces, bending

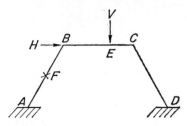

Figure 3.8. Lateral loading of a member

moments or displacements are required at some unloaded point along the member. For instance in Figure 3.8, a joint is introduced to the member *BC* at *E* and the portions *BE* and *EC* are treated as two members. If the deflection of the member at such a joint as *F* is required a further joint is also introduced there. In this manner the frame of Figure 3.8 has four joints *F*, *B*, *E*, and *C* that compose its stiffness matrix.

Distributed loading on a member can be treated as the sum of two systems of loads. These are

(i) The actual loading, together with end reactions and end moments that render the member fixed ended.

(ii) A set of end forces and end moments acting in the opposite direction to those of system (i). The structure is then analysed under the second system of loads and the results are superimposed with those of the first loading system. Superposition is possible provided that the analysis is carried out under a constant load factor. The actual loading and the two systems of loading of a portal frame are shown in Figure 3.9.

Members with thicker ends such as haunches and gusset plate can be treated accurately by introducing joints at the points where

112

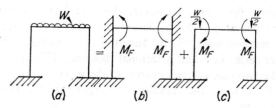

Figure 3.9. Equivalent forces on a frame
(a) Actual distributed load
(b) Loading system (i)
(c) Loading system (ii)

the sectional properties of a member change. The actual properties of the thick portions may then be used in the analysis. For instance for the member *ABCD* in Figure 3.10 two joints are introduced at *B* and *C* to divide it into three members *AB*, *BC* and *CD*.

Figure 3.10. A member with gusset plates

3.7. A COMPUTER ROUTINE FOR THE EVALUATION OF THE STABILITY FUNCTIONS

The procedure or routine discussed here can be incorporated in the main body of a programme for the non-linear analysis of structures. This routine is versatile in so far as it calculates the functions $\phi_1 - \phi_5$, the s, c, n, o and m functions as well as ϕ_6 and ϕ_7. However in a particular programme only some of these will be required. For this reason one of the formal parameters of the routine is an integer number i that specifies which set of these functions is needed. A second formal parameter is a real number r that specifies the value of ϱ for which the stability functions are required. The results are stored in an array of twelve locations, which is also declared as a formal parameter of the routine.

From the value of ϱ the main stability function ϕ_1 is calculated. The computer uses a power series approximation to calculate

9

trigonometric functions. However, the functions $\alpha \cot \alpha$ gives singular values at $\varrho = 4, 16, 36$ etc. For this reason Livesely[2] devised a method whereby this function is calculated as the sum of a power series in ϱ and a rational function. This arrangement absorbs the two singularities nearest to the working range $-4 \leqslant \varrho \leqslant 4$. In this manner the function converges rapidly. Thus ϕ_1 is calculated from

$$\phi_1 = \frac{64 - 60\varrho + 5\varrho^2}{(16 - \varrho)(4 - \varrho)} - \sum_{n=1}^{n=7} \frac{a_n \varrho^n}{2^{3n}} \qquad (3.36)$$

which gives ϕ_1 accurately to six decimal places, where

$$a_1 = 1 \cdot 57973627, \qquad a_2 = 0 \cdot 15858587, \qquad a_3 = 0 \cdot 02748899,$$

$$a_4 = 0 \cdot 00547540, \qquad a_5 = 0 \cdot 00115281, \qquad a_6 = 0 \cdot 00024908$$

and $\qquad\qquad\qquad\qquad a_7 = 0 \cdot 00005452.$

Once the value of ϕ_1 is obtained, the rest of the stability functions are readily calculated. A flow diagram of the routine is shown in Figure 3.11. It has been found that for very small values of ϱ, e.g. $\varrho < 0 \cdot 00001$, equation (3.36) leads to an exponential overflow and care has been taken in the programme to avoid this. As soon as the value of ϱ ($= r$) is transferred from the main programme to the routine it is tested and, if small, no further calculation is carried out. A caption indicating that the axial load is small is printed out. For larger values of r the stability functions ϕ_1 to ϕ_5 are calculated using equations (3.36) and (3.24).

The integer i may be given the values 1, 2, 3, or 4 and once ϕ_1 to ϕ_5 are calculated, the value of i decides what other stability functions are required. The value of $i = 1$ indicates that no further calculation is needed and the routine is left. The value of $i = 2$ indicates that s, c, m, n and o are also required and these are calculated using equations (3.21), (3.28) and (3.30). However if i is not equal to 2, equations (3.26) are used to calculate the values of ϕ_6 and ϕ_7. Having done this i is tested again and, if i is equal to 3, no further calculation is required. For $i = 4$ on the other hand s, c, m, n, and o are calculated. This indicates that if all the stability functions are required then i is set to 4. Once a set of functions is calculated, instructions may be given for their values to be printed.

114

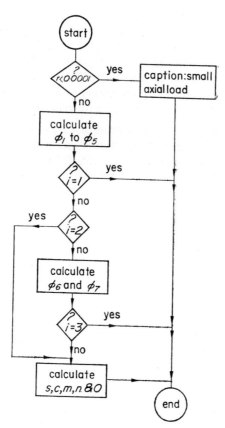

Figure 3.11. Flow diagram for a procedure to calculate stability functions

Exercises

1. Knowing that $s = 4\phi_3$, $c = \phi_4/2\phi_3$ show that $\pi^2 \varrho = 4(1 - mn)/m$.

2. *ABC* is a continuous beam pinned at *A* and *C* and simply supported at *B*. The beam is subject to an axial load *P*. Show that the rotation at *B* caused by an external moment acting there is given by

$$M/[s_1(1 - c_1^2)k_1 + S_2(1 - c_2^2)k_2]$$

where suffixes 1 and 2 refer to members *AB* and *BC* respectively.

3. In Figure 3.12 the member *ab* is connected to supports *AB* by gusset plates *aA* and *bB* that are assumed to be completely rigid. *B* is fixed while *A* is simply supported. The portion *ab* is of length *L* and has uniform flexural rigidity *EI*. Terminal moments m_A and m_B are induced by a rotation $\theta_A(=\theta_a)$ at *A*. The modified stability functions are thus given by $m_A = Sk\theta_A$ and $m_B = CSk\theta_A$, where $k = EI/L$. If $\varrho = PL^2/\pi^2EI$ derive expressions for *S* and *SC*.

(Note. These expressions were originally derived by Livesley and Chandler as a treatment of the effect of gusset plates.)

(Hint. Take moments about *A* for *Aa*, *B* for *Bb* and *A* for *AB*.)

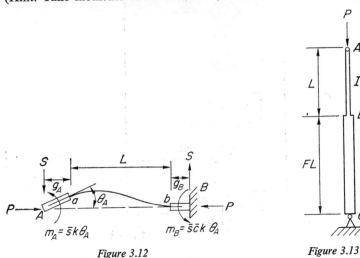

Figure 3.12

Figure 3.13

Answer. $$S = s + 2Ag_A(L+g_A)/L^2,$$
$$SC = sc + s(1+c)(g_A+g_B)/L + 2Ag_Ag_B/L^2$$

where $A = s(1+c) - \pi^2\varrho/2$, *s* and *c* are for portion *ab*.

4. The column *ABC* in Figure 3.13 is pinned at *A* and *C* and consists of two portions, *AB* of length *L* and second moment of area *I* and *BC* of length *FL* and second moment of area F^2L, where $F \neq 0$. The column is subject to an axial force *P* which causes point *B* to sway sideways. Using slope deflection method show that for any value of *P*, below the buckling load, point *B* does not rotate i.e. $\theta_B = 0$.

116

CHAPTER 4

Non-Linear elastic Frames

4.1. DERIVATION OF A NON-LINEAR STIFFNESS MATRIX

In the previous chapter the stiffness and flexibility equations for individual members were derived using various stability functions. These equations are utilised in this chapter to develop methods for the non-linear elastic analysis of structures. It is significant that, irrespective of the behaviour of the individual members composing a structure, the structure itself acts as an integral unit, where the stronger members and joints assist the weaker. The overall behaviour of the structure is therefore determined by the interaction between all its components.

In this chapter the stiffness matrix of a general member will be developed and this is then used in the matrix displacement method for the non-linear analysis of structures. In order to preserve versatility in the use of stability functions, the stiffness matrix of a member is derived using two different sets of these functions.

4.1.1. Using s and c functions

In Figure 4.1 a member of a rigidly jointed plane frame is shown in a general state of deformation. The member is subject to an axial force p, bending moments $M_{AB} = M_1$ and $M_{BA} = M_2$ as well as shear forces S as shown in the figure. These member forces cause an extension (or contraction) u, a sway v of the second end relative to its first end and rotations θ_{R1} and θ_{R2}. The general

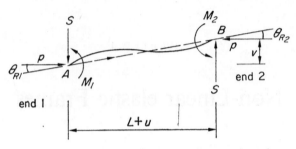

Figure 4.1. Forces and deformations of a general member

slope-deflection equations, taking the effect of axial forces into consideration, are given by equations (3.7) as

$$\left.\begin{array}{l} M_1 = -\dfrac{s(1+c)kv}{L} + sk\theta_1 + csk\theta_2 \\[3mm] M_2 = -\dfrac{s(1+c)kv}{L} + csk\theta_1 + sk\theta_2 \end{array}\right\} \tag{4.1}$$

The shear force in the member is given by the sum of equation (3.22), for pure sway and the last of equations (3.5) for pure rotation, thus

$$S = \frac{2s(1+c)kv}{mL^2} - \frac{s(1+c)k\theta_1}{L} - \frac{s(1+c)k\theta_2}{L} \tag{4.2}$$

The axial stiffness of the member is given by Hooke's law as

$$p = \frac{EA}{L}u \tag{4.3}$$

It should be pointed out that, in order to take the effect of the axial loads on bending into consideration, the axial loads themselves must be calculated at some stage and thus equation (4.3) has to be included in the analysis.

118

Writing equations (4.1) through (4.3) in matrix form

$$
\begin{bmatrix} p \\ S \\ M_1 \\ M_2 \end{bmatrix} = \begin{bmatrix} EA/L & 0 & 0 & 0 \\ 0 & \dfrac{2s(1+c)k}{mL^2} & -\dfrac{s(1+c)k}{L} & -\dfrac{s(1+c)k}{L} \\ 0 & -\dfrac{s(1+c)k}{L} & sk & csk \\ 0 & -\dfrac{s(1+c)k}{L} & csk & sk \end{bmatrix} \begin{bmatrix} u \\ v \\ \theta_1 \\ \theta_2 \end{bmatrix},
$$

$$(4.4)$$

or just:

$$P_i = k_i Z_i \qquad (4.5)$$

where i is a typical member of the frame. The matrix k is the stiffness matrix for the member.

4.1.2. Using ϕ functions

Using the relationships between the s and c functions and the ϕ functions as given in Section 3.4, it can be seen that

$$
\begin{aligned}
\frac{2s(1+c)k}{mL^2} &= \frac{2EI}{L^3} \cdot \frac{s(1+c)}{m} \\
&= \frac{2EI\phi_1}{L^3} \cdot 6\phi_2 \\
&= \frac{12EI\phi_5}{L^3} \\[4pt]
\frac{s(1+c)k}{L^2} &= \frac{EI}{L^2} s(1+c) \\
&= \frac{6EI}{L^2}\phi_2; \\[4pt]
sk &= \frac{EIs}{L} = \frac{4EI}{L}\phi_3. \\[4pt]
sck &= \frac{EI}{L}cs = \frac{2EI}{L}\phi_4
\end{aligned} \right\} \qquad (4.6)
$$

Substituting these in equations (4.4) results in

$$
\begin{bmatrix} p \\ S \\ M_1 \\ M_2 \end{bmatrix} = \begin{bmatrix} EA/L & 0 & 0 & 0 \\ 0 & \dfrac{12EI\phi_5}{L^3} & -\dfrac{6EI\phi_2}{L^2} & -\dfrac{6EI\phi_2}{L^2} \\ 0 & -\dfrac{6EI\phi_2}{L^2} & \dfrac{4EI\phi_3}{L} & \dfrac{2EI\phi_4}{L} \\ 0 & -\dfrac{6EI\phi_2}{L^2} & \dfrac{2EI\phi_4}{L} & \dfrac{4EI\phi_3}{L} \end{bmatrix} \begin{bmatrix} u \\ v \\ \theta_1 \\ \theta_2 \end{bmatrix} \quad (4.7)
$$

Denoting

$$
\left.\begin{aligned}
b &= \frac{12EI\phi_5}{L^3} \\
d &= -\frac{6EI\phi_2}{L^2} \\
e &= \frac{4EI\phi_3}{L} \\
f &= \frac{2EI\phi_4}{L}
\end{aligned}\right\} \quad (4.8)
$$

the stiffness matrix of equation (4.7) becomes

$$
k = \begin{bmatrix} a & \text{Symmetrical} & & \\ 0 & b & & \\ 0 & d & e & \\ 0 & d & f & e \end{bmatrix} \quad (4.9)
$$

It is immediately evident that these equations (4.7), (4.8) and (4.9) are very similar to equations (2.7), (2.9) and (2.10) for linear analysis. In fact we already know that in the absence of any axial forces, the ϕ functions all take the value of unity and equations (4.7), (4.8) and (4.9) are converted to (2.7), (2.9) and (2.10) respectively. This indicates how, fundamentally, the non-linear analysis of a structure is a continuation of the linear analysis. However the problem of non-linear analysis is complicated by the fact that to begin with, the axial loads in the members are themselves unknown. Previous to the development of matrix methods, analysis

120

avoided this difficulty by either assuming the values of the axial loads or by calculating them after making a number of assumptions that simplified the actual problem. Later in this chapter, methods are given for the non-linear analysis of structures that make use of these simplifications.

4.2. MATRIX DISPLACEMENT METHOD FOR NON-LINEAR ANALYSIS OF STRUCTURES

With the aid of the stiffness matrix as given by equations (4.4), (4.7) or (4.9) the matrix displacement method for the non-linear analysis of structures can be formulated. Before doing this, it should be pointed out that, while the stiffness matrix of a member is influenced by the presence of an axial force, the displacement transformation matrix A of the structure is unaltered. This is because the displacement transformation matrix relates the displacement of the members, after the axial loads have played their part, to the displacements of the joints in the structure.

As mentioned earlier, the axial forces in the members are to begin with, unknown and therefore the stability functions for the members cannot be calculated. To overcome this difficulty an iterative method is employed, whereby the axial forces are initially neglected and a linear analysis is performed to estimate the joint displacements and then the member forces. These member forces are then used to calculate the stability functions for the members so that these may be used to construct an improved stiffness matrix. This enables a reanalysis of the structure to be made under the same set of externally applied loads. This iteration can be continued until the difference between the axial loads of the members, used to calculate the stability functions, and those obtained at the end of the same cycle, is within a specific tolerance.

The steps required for a complete non-linear elastic analysis of a structure can be summarised as follows

1. Select a load factor λ for the external load system and assume the axial loads in the members to be zero.
2. Construct the overall stiffness matrix K either directly or by the triple multiplication $A.k.A$.

3. Solve the joint equilibrium equations $L = K.X$ for the joint displacements X.
4. Construct the product kA either directly or use those obtained in step 2 and calculate the member forces using $P = k.A.X$.
5. Use the axial loads in the members and calculate their stability functions.
6. Repeat the process from step 2 until the difference between two successive sets of axial loads is smaller than a specific tolerance.
7. Repeat the process from step 1 with a new load factor.

In this manner an accurate set of joint displacements and member forces become available for each load factor and the non-linear load-displacement diagram of the structure can be plotted.

The overall stiffness matrix K of the structure is constructed using the same procedure as that for linear analysis and the notations used here will be the same as those adopted in Chapter 2.

4.3. WORKED EXAMPLES

As an application of the Matrix displacement method a number of examples are solved. In particular the first example uses a simple cantilever to point out characteristic features common to nonlinear behaviour of all structures. A newcomer to the field would be well advised to remember these characteristics when considering the non-linear behaviour of more complex structures.

The cantilever column of Figure 4.2 is 300 mm long and its cross-section is 10 mm×10 mm with a modulus of elasticity

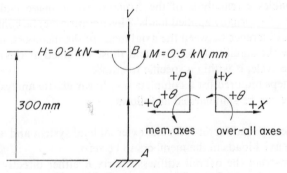

Figure 4.2. Forces on a cantilever column

$E = 207$ kN/mm². At end B the column is subject to a horizontal force of 0·2 kN and a moment of 0·5 kN.mm as shown in the figure. It is required to calculate the slopes and the horizontal displacements of the free end when the column is also subject to vertical forces of zero, 3·8 kN, 4·73 kN and 8·83 kN.

The axial force p in this column is known by vertical equilibrium to be equal to the applied vertical force V and therefore there is no need for an iterative process to assess it.

With $V = p = 0$ the overall stiffness matrix of the column is constructed for values of ϕ_1 to ϕ_5 all being unity. Adopting the same notations of equations (2.9) the stiffness matrix of the member will be given by equation (2.10), which is reproduced here for convenience.

$$k_{AB} = \begin{bmatrix} a & 0 & 0 & 0 \\ 0 & b & d & d \\ 0 & d & e & f \\ 0 & d & f & e \end{bmatrix}$$

The displacement transformation matrix is given by $Z = A.X$ i.e.

$$\begin{bmatrix} u \\ v \\ \theta_{AB} \\ \theta_{BA} \end{bmatrix} = \begin{bmatrix} 0 & 1 & 0 \\ -1 & 0 & 0 \\ 0 & 0 & 0 \\ 0 & 0 & 1 \end{bmatrix} \cdot \begin{bmatrix} x_B \\ y_B \\ \theta_B \end{bmatrix}$$

Thus $L = KX$ becomes:

$$\begin{bmatrix} -0·2 \\ 0 \\ 0·5 \end{bmatrix} = \begin{bmatrix} b & 0 & -d \\ 0 & a & 0 \\ -d & 0 & e \end{bmatrix} \begin{bmatrix} x_B \\ y_B \\ \theta_B \end{bmatrix} \qquad (4.10)$$

Equations (4.10) can also be used for the case of the non-linear analysis provided it is understood that the quantities b, d and e are in this case given by equations (4.8).

Solving equations (4.10) for θ_B and x_B gives

$$\theta_B = \frac{0·5b - 0·2d}{eb - d^2} \left.\begin{array}{c} \\ \\ \end{array}\right\}$$

and

$$x_B = \frac{d\theta_B - 0·2}{b} \qquad (4.11)$$

123

We have $\quad EI = \dfrac{207 \times 10 \times 1000}{12} = 172\,500 \ \text{kN mm}^2$

$$b = \frac{12EI}{L^3}\,\phi_5 = 0\!\cdot\!077\,\phi_5$$

$$d = -\frac{6EI}{L^2}\,\phi_2 = -11\!\cdot\!5\,\phi_2$$

$$e = \frac{4EI}{L}\,\phi_3 = 2300\,\phi_3$$

(i) with $\quad V = 0, \qquad \varrho = 0, \qquad \phi_2 = \phi_3 = \phi_5 = 1$

$$eb - d^2 = 44$$

Substituting these figures in equations (4.11) We obtain

$$\theta_B = 0\!\cdot\!053 \ \text{radians};$$
$$x_B = -10 \ \text{mm}$$

The negative sign indicates that the sway is to the left

(ii) With $\qquad V = p = 3\!\cdot\!8 \ \text{kN}, \qquad \varrho = p/P_E = \dfrac{3\!\cdot\!8 \times 300 \times 300}{172\,500\,\pi^2}$

i.e. $\qquad\qquad\qquad\qquad \varrho = 0\!\cdot\!2$

From the tables in Appendix 1 we find

$$\phi_2 = 0\!\cdot\!9666$$
$$\phi_3 = 0\!\cdot\!9324$$
$$\phi_5 = 0\!\cdot\!802$$

Hence, $\qquad\qquad b = 0\!\cdot\!077 \times 0\!\cdot\!802 = 0\!\cdot\!0615$

$$d = -11\!\cdot\!105; \qquad e = 2142$$

and $\qquad\qquad\qquad eb - d^2 = 8\!\cdot\!5$

Hence $\qquad \theta_B = 0\!\cdot\!278 \ \text{rad}. \qquad x_B = -53 \ \text{mm}$

(iii) With $\qquad V = p = 4\!\cdot\!73, \qquad \varrho = \dfrac{p}{P_E} = 0\!\cdot\!25$

124

Before proceeding to calculate θ_B and x_B, we notice that

$$eb - d^2 = \frac{4EI}{L}\phi_3 \times \frac{12EI}{L^3}\phi_5 - \frac{36(EI)^2}{L^4}\phi_2^2,$$

and since $\qquad \phi_5 = \phi_1\phi_2$

$$\therefore \; eb - d^2 = \frac{12(EI)^2}{L^4}\phi_2(4\phi_3\phi_1 - 3\phi_2)$$

with $\qquad \varrho = 0{\cdot}25, \qquad \phi_1 = 0{\cdot}78542,$

$\qquad\qquad\quad \phi_2 = 0{\cdot}95811, \qquad \phi_3 = 0{\cdot}91494$

and $\quad 4\phi_3\phi_1 - 3\phi_2 = 4 \times 0{\cdot}91494 \times {\cdot}78542 - 3 \times 0{\cdot}95811 = 0$

$\therefore$ at $\qquad \varrho = 0{\cdot}25, \qquad eb - d^2 = 0, \qquad \theta_B = -x_B = \infty$

(iv) With $\qquad V = p = 8{\cdot}83, \qquad \varrho = 0{\cdot}47$

$\qquad eb - d^2 = -39{\cdot}5, \qquad \theta_B = -0{\cdot}54$ rad, $\qquad x_B = 10$ mm

The results of these analyses are shown in Table 4.1.

Table **4.1** SUMMARY OF RESULTS

$V = p$	$\varrho = p/P_E$	$eb - d^2$	θ_B	x_B
0	0	44	0·053	−10
3·8	0·2	8·5	0·278	−53
4·73	0·25	0	∞	−∞
8·83	0·47	−39·5	−0·54	10

From these results the following conclusions can be drawn

(i) That the value of θ_B depends not only on ϱ but also on the quantity $eb - d^2$. As ϱ increases from 0 to 0·25, the value of $eb - d^2$ decreases from 44·0 to zero. After that $eb - d^2$ becomes negative; so does the value of θ_B. The horizontal sway also changes its direction.

(ii) When $p = 4{\cdot}73$, the deflections become infinitely large. At this value $eb - d^2$ vanishes. In fact the force $4{\cdot}73 = \pi^2 EI/(4L^2)$ is the Euler critical load for a column which is fixed at one end and free at the other.

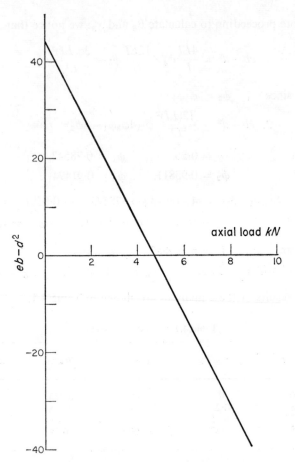

Figure 4.3. Graph of $eb - d^2$ against p

(iii) A graph of $(eb - d^2)$ against the axial load p is shown in Figure 4.3, which is seen to be nearly a straight line.

(iv) It is observed from the overall stiffness matrix of the structure, given in equation (4.10), that the determinant of this matrix is

$$\text{Dit.} = \begin{vmatrix} b & 0 & -d \\ 0 & a & 0 \\ -d & 0 & e \end{vmatrix} = a(eb - d^2),$$

126

when $eb - d^2$ vanishes, the value of the determinant of the stiffness matrix becomes zero while displacements increase infinitely, indicating that at this point the three rows of the stiffness matrix are not independent. This state of zero determinant is known as the state of elastic instability and will be looked at in detail in the next chapter.

As a second example the pitched roof frame of Figure 4.4 is analysed in order to obtain its non-linear load-displacement

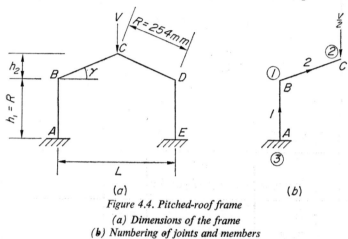

Figure 4.4. Pitched-roof frame
(a) Dimensions of the frame
(b) Numbering of joints and members

diagram. As shown in the figure the frame is symmetrical with a point load acting vertically at its apex. The angle of pitch γ is $22\frac{1}{2}°$ and the members are of equal length and constant cross section throughout with the width b being 13 mm and the depth d as 3 mm. Because of symmetry only half the frame is analysed and the numbering of members and joints for this half is shown in Figure 4.4(b). The axial loads in members 1 and 2 are different and therefore their ϱ values are also different.

The axial stiffness of the members is taken into consideration and thus joint 1 has three degrees of freedom in X, Y and θ directions, but because of symmetry joint 2 can only move vertically. Thus the joint displacement vector is $x = \{x_1 \ y_1 \ \theta_1 \ y_2\}$. The corresponding load vector is

$$L = \{H_1 \ V_1 \ M_1 \ V_2\} = \{0 \ 0 \ 0 \ 0.5V\}$$

Using the procedure described in Section 2.4 for the construction of the overall stiffness matrix as summarised by equations (2.41)

127

and (2.42) and expressing the direction cosines of member 2 in terms of the angle of pitch, the overall stiffness equations $L = KX$ for the half-frame becomes

$$
\begin{bmatrix} 0 \\ 0 \\ 0 \\ \dfrac{V}{2} \end{bmatrix} =
$$

$$
\begin{bmatrix}
b_1 + a_2\cos^2\gamma + b_2\sin^2\gamma & & \text{Symmetrical} & \\
(a_2-b_2)\sin\gamma\cos\gamma & a_1 + a_2\sin^2\gamma + b_2\cos^2\gamma & & \\
-d_1 + d_2\sin\gamma & -d_2\cos\gamma & e_1 + e_2 & \\
(b_2-a_2)\sin\gamma\cos\gamma & -a_2\sin^2\gamma - b_2\cos^2\gamma & d_2\cos\gamma & a_2\sin^2\gamma + b_2\cos^2\gamma
\end{bmatrix}
$$

$$
\times \begin{bmatrix} x_1 \\ y_1 \\ \theta_1 \\ y_2 \end{bmatrix} \qquad (4.12)
$$

where the contribution of members 1 and 2 to the overall stiffness matrix is distinguished by suffixes 1 and 2 respectively. The quantities a, b, d and e are those defined by equations (4.8).

128

For a linear analysis $\phi_1 - \phi_5$ are all unity and the solution of equations (4.12) for $V = -0.267$ kN gives the joint displacement vector X as:

$$X = \{-9.16 \quad -0.0042 \quad -1.11 \quad -22.12\} \text{ mm}$$

These displacements are used with $P = k.A.X$ and the member forces are obtained to be

$$p_1 = 0.133 \text{ kN compressive}$$

and $p_2 = 0.121$ kN compressive

Accordingly, the stability functions are found from the tables in Appendix 1. These values are used to construct the overall stiffness matrix once again and equations (4.12) are solved with the same value of V. The resulting joint displacements are utilised to obtain improved values for the member forces. This process was repeated three times to obtain the final set of displacements. These are:

$$X = \{-10.58 \quad -0.0042 \quad -1.30 \quad -25.81\} \text{mm}$$

Comparing these with the results of the linear analysis shows the small difference that exists between the linear and non-linear displacements. This was the reason for selecting such a slender frame for the analysis. However, once the load factor increases, the divergence between the load displacement diagrams by the two methods becomes more apparent. The results also show that the vertical displacement y_1 of joint 1 is very small compared to the other elements of the displacement vector. Neglecting the axial stiffness of the members, therefore, would not have caused an appreciable difference to the analysis at this stage.

The iteration was then repeated with a vertical load of 0.311, 0.356, 0.40, 0.533 and 0.667 kN acting downwards at the apex and the load displacement curves for the linear and the non-linear analyses are shown in Figure 4.5. On the same figure the variation of the axial loads with the load factor is also shown where it can be seen that the axial loads in the members are nearly linearly related to the load factor. This indicates that once these axial loads are obtained at a given load factor, such as say $V = 0.267$ kN the axial loads for any other load factor can be readily obtained with reasonable accuracy.

It is noticed from the load-deflection graph of Figure 4.5 that at a load of 0.667 kN at the apex, the vertical deflection there is

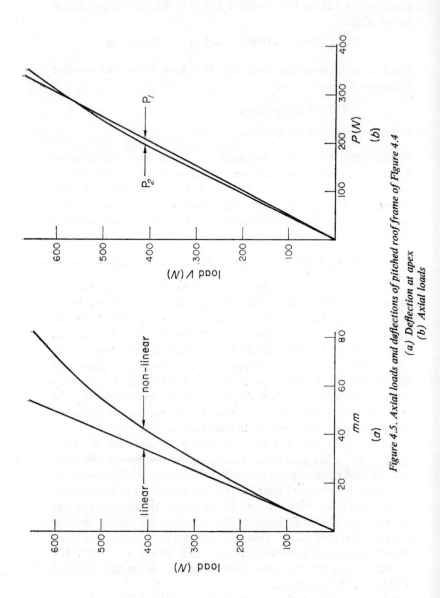

Figure 4.5. Axial loads and deflections of pitched roof frame of Figure 4.4
(a) Deflection at apex
(b) Axial loads

87·6 mm which is rather large. It is doubtful whether the small deflection theory is applicable in this case. It took six iterations for the analysis to converge at this load. At higher loads it was not possible to make the iteration converge with as many as fifteen cycles.

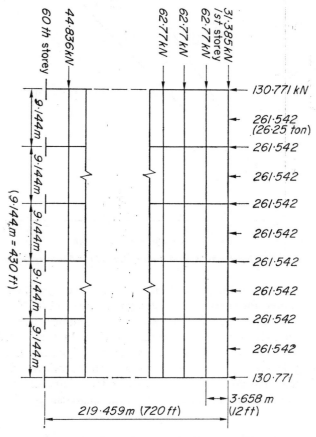

Figure 4.6. 60-storey 5-bay frame dimensions and loading. All beams are similarly loaded

As a final example the sixty-storey five-bay frame of Figure 4.6 is briefly discussed. The frame is 219·456 m high with equal bays and storey heights. It consists of 960 members and 665 joints of which five are rigidly connected to the supports and therefore the frame has 1980 degrees of freedom. A computer programme, that fol-

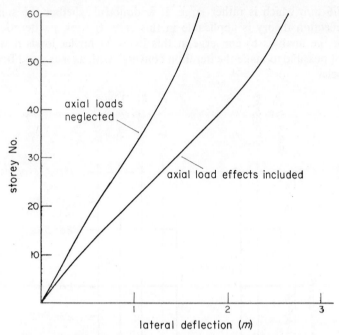

Figure 4.7. *Deflections at various storey levels for a 60-storey frame*

lowed the steps in the last section, was used to analyse the frame. This required four iterations to satisfy a tolerance of 0·5% on the axial loads calculated by two successive solutions. The whole operation took 9 minutes of computer time. In Figure 4.7 the deflection at each storey level is plotted both ignoring and considering the effect of axial loads. It is noticed that when taking the axial loads into consideration, the deflection of the top storey is 40% more than that obtained by the linear analysis.

4.4. TRIANGULATED FRAMES

The matrix displacement method for the non-linear structures discussed in Section 4.2 can be simplified considerably when the axial loads in the members can be estimated. Even though the assumptions involved in this estimation render the initial analysis approximate a knowledge of the axial loads can speed up the convergence of the iteration process.

132

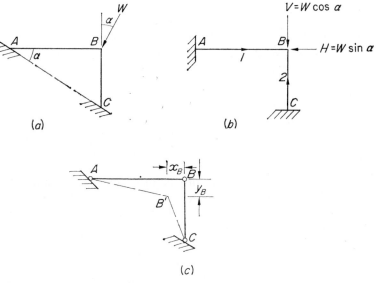

Figure 4.8. Triangulated frame
(a) Frame and loading
(b) Applied load vector and numbering of members
(c) Deformations of the determinate pin-jointed frame

Some rigidly-jointed triangulated frames can be rendered statically determinate by inserting pins at their joints. The axial loads
in these frames can be calculated reasonably accurately by analysing
the resulting pin-jointed frame. These axial forces can be utilised
in the study of secondary stress distribution in the original rigidly
jointed frame. Consider the simple triangular frame ABC of Figure
4.8. The frame is fixed at A and C and rigidly jointed at B. A force
W is acting at right-angles to the line joining A and C. This force
is resolved in Figure 4.8(b) into its components V and H.

The joint equilibrium equation $L = KX$ for this frame is constructed in the usual manner, thus

$$\begin{bmatrix} H \\ V \\ \hline 0 \end{bmatrix} = \begin{bmatrix} a_1+b_2 & 0 & -d_2 \\ 0 & b_1+a_2 & d_1 \\ \hline -d_2 & d_1 & e_1+e_2 \end{bmatrix} \begin{bmatrix} x_B \\ y_B \\ \hline \theta_B \end{bmatrix} \qquad (4.13)$$

where the suffixes 1 and 2 refer to the member numbers.

133

This frame can be rendered statically determinate by inserting three hinges at A, B and C as shown in Figure 4.8(c). The axial forces in the members of the new frame can be calculated from statics as: $p_{AB} = H$ and $p_{BC} = V$, while the shortening of the members is obtained by Hooke's law as

$$u_{AB} = -\frac{HL_1}{E_1A_1} = -\frac{WL_1 \sin \alpha}{E_1A_1},$$

$$u_{BC} = -\frac{VL_2}{E_2A_2} = -\frac{WL_2 \cos \alpha}{E_2A_2}$$

It follows directly from the displacement transformation matrix that $x_B = u_{AB}$ and $y_B = u_{BC}$ Hence in equations (4.13) we have only one unknown, θ_B, the value of which can best be calculated using matrix partitioning.

In general a set of equations $L = KX$ can be partitioned as

$$\begin{bmatrix} L_1 \\ L_2 \end{bmatrix} = \begin{bmatrix} a_{11} & a_{12} \\ a_{21} & a_{22} \end{bmatrix} \begin{bmatrix} X_1 \\ X_2 \end{bmatrix} \tag{4.14}$$

where either X_1 or X_2 may be a known vector. If for instance X_1 are known while X_2 are unknowns then the second of equations (4.14) gives

$$L_2 = a_{21} X_1 + a_{22} X_2$$

Thus

$$X_2 = a_{22}^{-1} \cdot (L_2 - a_{21} X_1) \tag{4.15}$$

which gives the values of X_2 in terms of the known matrices of the right-hand side. The technique used here is general and in our particular case equations (4.13) can be partitioned so that

$$X_2 = \theta_B,$$

while

$$X_1 = \{x_B \quad y_B\}$$

$$a_{22} = e_1 + e_2$$

$$a_{21} = [-d_2 \quad d_1]$$

Using equations (4.15), it follows that

$$\theta_B = (e_1 + e_2)^{-1}(O - [-d_2 \quad d_1]\{x_B \quad y_B\})$$

134

i.e.
$$\theta_B = \frac{d_2 x_B - d_1 y_B}{e_1 + e_2} \tag{4.16}$$

Once all the elements of the joint displacement vector are known, the member forces throughout the structure are obtained as usual from $P = k \cdot A \cdot X$ of the original rigidly jointed structure. For the special case of $\alpha = 30°$, $L = L_1$, $E_1 I_1 = E_2 I_2 = EI$, $A_1 = A_2 = A$, for instance, equation (4.16) becomes

$$\theta_B = \frac{3W}{4EA} \left[\frac{3\phi_{2,2} - \phi_{2,1}}{\phi_{3,1} + \sqrt{3}\phi_{3,2}} \right] \tag{4.17}$$

where the second suffix of the stability functions refers to the member number. The moments M_{BA} and M_{BC} are therefore:

$$
\left.
\begin{aligned}
M_{BA} &= d_1\, y_B + e_1\, \theta_B \\
&= \frac{3WI\phi_{2,1}}{AL} + \frac{4EI\theta_B}{L}\, \phi_{3,1} \\[2mm]
\text{and} \quad M_{BC} &= -d_2\, x_B + e_2\, \theta_B \\
&= \frac{9WI\phi_{2,2}}{AL} + \frac{4\sqrt{3}EI}{L}\, \theta_B\, \phi_{3,2}
\end{aligned}
\right\} \tag{4.18}
$$

Alternatively, once the axial loads in the members are estimated from the pin-jointed frame, the matrix displacement method of Section 4.2 can be entered at step 5 by using the estimated forces to calculate the stability functions and proceeding to step 6 as usual. This helps the iteration process to converge faster.

As an example consider the rigidly-jointed triangulated frame of Figure 4.9 which is fixed to the supports at C and D. The dimensions and the numbering of the joints and the members are shown in Figure 4.9(a). All the members have the same constant cross section and the frame is subject to a horizontal force W at A and another force W at B which acts at 60 degrees to the vertical.

This frame is rendered statically determinate by inserting hinges at all the joints. The axial forces and extensions of each member of the resulting pin-jointed structure are calculated by statics as given in Table 4.2 with compressive forces and member shortenings shown positive.

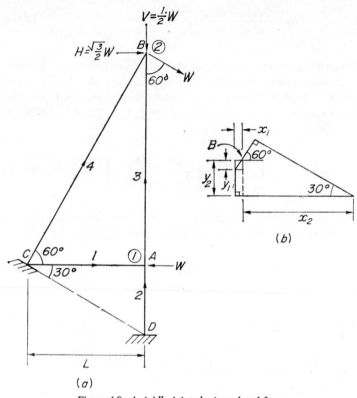

Figure 4.9. A rigidly-jointed triangulated frame
(a) loads and dimensions
(b) Williot Mohr diagram for the frame

Table 4.2 MEMBER FORCES AND EXTENSIONS OF THE FRAME, SHOWN IN FIGURE 4.9

Member	Length	Force	Contraction	ϕ_2	ϕ_3	ϕ_4^2
1	L	W	$\dfrac{WL}{EA}$	0·9496	0·8972	
2	$L/\sqrt{3}$	2W	$\dfrac{2WL}{\sqrt{3}EA}$	0·9666	0·9324	
3	$\sqrt{3}L$	2W	$\dfrac{2\sqrt{3}WL}{EA}$	0·6569	0·1793	2·5992
4	2L	$\sqrt{3}W$	$\dfrac{-2\sqrt{3}WL}{EA}$	1·3012	1·5546	

136

The equilibrium equations $L = KX$ can be constructed in the usual manner and is given below using the same notations of equations (4.14) and (4.15), with rows and columns corresponding to θ_1 and θ_2 appearing together:

$$\begin{bmatrix} L_1 \\ L_2 \end{bmatrix} = \begin{bmatrix} a_{11} & a_{12} \\ a_{21} & a_{22} \end{bmatrix} \begin{bmatrix} X_1 \\ X_2 \end{bmatrix}$$

where

$$L_1 = \{H_1 \ V_1 \ H_2 \ V_2\},$$
$$L_2\{M_2 \ M_1\},$$
$$X_1\{x_1 \ y_1 \ x_2 \ y_2\},$$
$$X_2\{\theta_2 \ \theta_1\},$$

$$a_{11} = \begin{bmatrix} a_1+b_2+b_3 & 0 & -b_3 & 0 \\ 0 & b_1+a_2+a_3 & 0 & -a_3 \\ -b_3 & 0 & b_3+\dfrac{a_4+3b_4}{4} & \dfrac{\sqrt{3}a_4}{4}-\dfrac{\sqrt{3}b_4}{4} \\ 0 & -a_3 & \dfrac{\sqrt{3}(a_4-b_4)}{4} & a_3+3a_4+b_4 \end{bmatrix};$$

$$a_{12} = \begin{bmatrix} d_3 & d_3-d_2 \\ 0 & d_1 \\ -d_3-\dfrac{\sqrt{3}d_4}{2} & -d_3 \\ \dfrac{d_4}{2} & 0 \end{bmatrix}$$

$$a_{21} = \begin{bmatrix} d_3 & 0 & -d_3-\dfrac{\sqrt{3}d_4}{2} & \dfrac{d_4}{2} \\ -d_2+d_3 & d_1 & -d_3 & 0 \end{bmatrix}$$

$$a_{22} = \begin{bmatrix} e_3+e_4 & f_3 \\ f_3 & e_1+e_2+e_3 \end{bmatrix}$$

In Figure 4.9(b) the joint translations X_1 are obtained from the member extensions using a Williot Mohr diagram. These are

$$X_1 = \{x_1 \ y_1 \ x_2 \ y_2\} = \frac{WL}{EA}\left\{-1 \ \frac{2}{\sqrt{3}} \ 4(2+\sqrt{3}) \ -8/\sqrt{3}\right\}$$

From Table 4.2, the ratio of $\varrho = p/P_E$ for the members can be calculated in terms of ϱ_1. These are $\varrho_2 = \frac{2}{3}\varrho_1$, $\varrho_3 = 6\varrho_1$ and $\varrho_4 = -4\sqrt{3}\varrho_1$ and for a value of $\varrho_1 = 0\cdot3$ for instance, $W = 0\cdot3\pi^2EI/L^2$ We obtain $\varrho_2 = 0\cdot2$, $\varrho_3 = 1\cdot8$ and $\varrho_4 = -2\cdot08$. The stability functions can now be obtained from the tables in Appendix 1 and these are also shown in Table 4.2. Substituting these in equation (4.15) with the above matrices and solving for θ_1 and θ_2, for an axial strain W/EA of 10^{-5} we obtain

$$\theta_1 = -27\cdot67\times10^{-5} \quad \text{rad}$$
$$\theta_2 = 4\cdot08\times10^{-5} \quad \text{rad}$$

With all the elements of the joint displacement vector now known, the bending moments can be calculated.

Before leaving this example, it is interesting to mention that the joint translations of the pin-jointed frame were obtained using the Williot Mohr diagram of Figure 4.9(b). However when dealing with complex structures, this diagram becomes complicated and it is rather paradoxical to construct such a diagram before using a computer for the rest of the analysis. When dealing with matrix force method, however, we showed that

$$L'X = P'Z$$

and since $P = BL$ i.e. $P' = L'B'$, it follows that

$$L'X = L'B'Z$$

Hence $\qquad\qquad\qquad X = B'Z$

This means that using the tranpose of the load transformation matrix B, the joint displacements can be calculated from the member displacements which is precisely what the Williot Mohr diagram does. It should be pointed out, however, that when B is constructed the actual load should be replaced with H and V at each joint in the structure so that the load vector L has the same number of elements as the joint displacement vector.

4.5. INITIAL STRESSES

During the manufacturing or erection of structures it is common to introduce initial internal stresses in them due to the lack of fit of the members. In Figure 4.10 some examples of lack of fit are

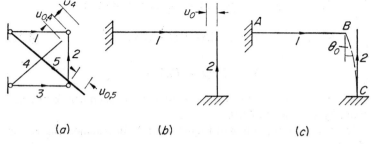

(a)   (b)   (c)

Figure 4.10. Structures with initial stresses
(a) member 4 too short by $u_{0,4}$. Member 5 too long by $u_{9,5}$
(b) member 1 too short by u_0
(c) angular lack of fit of θ_0

given, where u_0 is the amount by which one or more members are short. On the other hand θ_0 is a rotational lack of fit as shown by Figure 4.10(c) where due to fixing of supports A and C before joining AB and BC together, a moment may be required to be applied at B to make the members meet for fixing.

The lack of fit of various members of a structure can be exppssɔɹ in the form of a vector Z_0. For instance for the pin-jointed frame of Figure 4.10(a), this vector is

$$Z_0 = \{u_{0,1} \ u_{0,2} \ u_{0,3} \ u_{0,4} \ u_{0,5}\} = \{0 \ 0 \ 0 \ u_{0,4} - u_{0,5}\}$$

indicating that there is no lack of fit in the first three members. Similarly the lack of fit vector for the frame of Figure 4.10(b) is

$$Z_0 = \{u_{AB} \ v_{AB} \ \theta_{AB} \ \theta_{BA} \ u_{CB} \ v_{CB} \ \theta_{CB} \ \theta_{BC}\}_0 = \{u_{0,1} \ 0 \ 0 \ 0 \ 0 \ 0 \ 0 \ 0\}$$

If the member displacements after closing the gaps and applying a set of forces are given by a vector $Z + Z_0$, it follows that the member forces P are given by:

$$P = k(Z + Z_0)$$
but
$$Z = AX$$
∴
$$P = kAX + kZ_0 \tag{4.19}$$

Now the equilibrium, between these forces and the externally applied loads L, is given by $L = A'P$; it follows using (4.19) that

$$L = A'kAX + A'kZ_0 \tag{4.20}$$

139

From these the joint displacements are calculated as

$$X = (A'kA)^{-1}\cdot(L - A'kZ_0) \qquad (4.21)$$

i.e.

$$X = K^{-1}(L - A'kZ_0)$$

Once these displacements are calculated the member forces are given by equations (4.19).

In the case of the frame of Figure 4.10(c), for instance, the vector Z_0 is:

$$Z_0 = \{0\ 0\ 0\ 0\ 0\ 0\ 0\ \theta_0\},$$

the relationship $Z = A.X$ is

$$
\begin{bmatrix} u_{AB} \\ v_{AB} \\ \theta_{AB} \\ \theta_{BA} \\ u_{CB} \\ v_{CB} \\ \theta_{CB} \\ \theta_{BC} \end{bmatrix}
=
\begin{bmatrix}
1 & 0 & 0 \\
0 & 1 & 0 \\
0 & 0 & 0 \\
0 & 0 & 1 \\
0 & 1 & 0 \\
-1 & 0 & 0 \\
0 & 0 & 0 \\
0 & 0 & 1
\end{bmatrix}
\begin{bmatrix} x_B \\ y_B \\ \theta_B \end{bmatrix}
$$

Matrix k can be readily constructed using equations (4.9) and $A'kZ_0$ is then given by

$$
A'kZ_0 =
\begin{bmatrix}
a_1 & 0 & 0 & 0 & 0 & -b_2 & -d_2 & -d_2 \\
0 & b_1 & d_1 & d_1 & a_2 & 0 & 0 & 0 \\
0 & d_1 & f_1 & e_1 & 0 & d_2 & f_2 & e_2
\end{bmatrix}
\begin{bmatrix} 0 \\ 0 \\ 0 \\ 0 \\ 0 \\ 0 \\ 0 \\ \theta_0 \end{bmatrix}
$$

$$
=
\begin{bmatrix} -d_2\theta_0 \\ 0 \\ e_2\theta_0 \end{bmatrix}
$$

140

From equations (4.20) or (4.21), substituting for

$$L = \{H_B \quad V_B \quad M_B\},$$

these being the externally applied loads at B, it follows that

$$\left[\begin{array}{cc|c} a_1+b_2 & 0 & -d_2 \\ 0 & b_1+a_2 & d_1 \\ \hline -d_2 & d_1 & e_1+e_2 \end{array}\right] \left[\begin{array}{c} x_B \\ y_B \\ \hline \theta_B \end{array}\right] = \left[\begin{array}{c} H_B+d_2\theta_0 \\ V_B \\ \hline M_B-e_2\theta_0 \end{array}\right] \quad (4.22)$$

Comparing equations (4.22) with (4.13), it is immediately evident that the partitioning technique of equations (4.14) and (4.15) can be usefully employed. In the present case, using the same notations of (4.14), we have

$$X_2 = \theta_B,$$
$$X_1 = \{x_B \quad y_B\},$$
$$a_{22} = e_1+e_2,$$
$$a_{21} = [-d_2 \quad d_1],$$
$$L_2 = M_B-e_2\theta_0$$

and
$$L_1 = \{H_B+d_2\theta_0 \quad V_B\}$$

Thus using equations (4.15) we obtain

$$\theta_B = \frac{M_B-e_2\theta_0+d_2x_B-d_1y_B}{e_1+e_2} \quad (4.23)$$

which is analogous to equation (4.16). The axial forces and joint translations here cannot be calculated by first inserting hinges at the joints. This is because with the existence of a gap the two members act as a mechanism and can be moved about until they meet without inducing any initial forces in the members of the resulting pin jointed frame. This difficulty can be overcome by first carrying out a linear analysis of the rigidly jointed frame, in order to estimate the axial forces in the members. The stability functions and the joint translations can then be calculated and used in equation (4.23) to evaluate θ_B. The rest of the problem is straightforward.

4.6. A COMPUTER PROGRAMME FOR THE NON-LINEAR ANALYSIS OF PLANE FRAMES

When presenting frame problems to a computer it is first necessary to construct a schematic diagram of the frame to show the numbering of the joints and the members. Each joint in this diagram is numbered from 1 upwards except for the supports. The

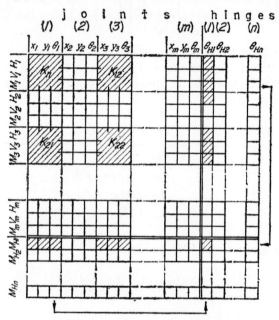

Figure 4.11. Layout of overall stiffness matrix showing the contribution of a member joining joint (1) and (3) with a hinge at end (1)

joint number for all the supports is zero indicating that in the overall stiffness matrix there will be no row or column to correspond to these supports. Each real hinge in the frame is also numbered. In Figure 4.11 a layout of the overall stiffness matrix is shown. In this figure there are three rows and three columns in this matrix corresponding to each joint; These are followed by rows and columns corresponding to the hinges. For this reason the first hinge is numbered as $3m+1$, where m is the total number of joints in the frame. The next hinge is numbered $3m+2$ and so on.

It is evident that a hinge number specifies the row and column belonging to that hinge.

The members of the frame are numbered from one upwards. and an arrow is placed against each member to identify its first end from its second end. The head of the arrow points to the second end of the member. The order in which the joints, hinges and members are numbered is arbitrary. In Figure 4.12 the schematic diagram of a frame is shown where it is noticed that joint B

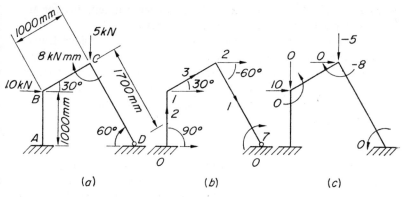

Figure 4.12. Schematic diagram and joint loads for a frame

(a) Frame and loading
(b) Schematic diagram
(c) Joint loads

is numbered as the first joint and joint C as the second. Joints A and D are both supports and thus they are numbered zero. There are two joints in this frame and thus $m = 2$ and the hinge at D is therefore numbered $3m+1 = 7$.

The information required to carry out a non-linear analysis of a frame is stored by means of the frame data. This consists of the modulus of elasticity E, of the members, the total number of members N, joints m, hinges n and the total number W of load factors to be considered. The data required to specify each member consists of the area A, second moment of area I, length L, inclination α, joint numbers J_1 and J_2 at its first and second end respectively and hinge numbers h_1 and h_2 at each end of the member. A zero hinge number at an end of a member indicates that there is no hinge at that end.

The externally applied load vector is presented to the computer in the form of an array L which is of order $3m+n$. For each joint there are three elements in the applied load vector. These are the horizontal load, the vertical load and the externally applied moment at the joint. The last n elements of the load vector correspond to the bending moments that can be transmitted across the hinges. In the case of real hinges these elements are zeros. In Figure 4.12(c) the elements of the load vector for the frame of Figure 4.12(a) are shown and Table 4.3 contains the data required by the computer for the analysis of this frame for one load factor. The last number in the data indicates the tolerance t acceptable for the convergence criterion of the iteration process.

The programme is for the non-linear analysis of plane frames and therefore use can be made of the special features of these frames. For instance the direction cosines are replaced by the sine and cosine of the angles of inclination of the members. Furthermore instead of joint coordinates and degrees of freedom of the joints, the data for the construction of the stiffness matrix are given in the form of member length, inclination and a statement as to whether or not there is a hinge at the end of a member.

The stiffness matrix is constructed by starting with a null matrix and then adding in the contributions for each member in their numerical order. These contributions are added directly to the locatio ns corresponding to the joints. This method avoids scanning operat ions through the joints and the members to find the members connected to each joint. If a member AB connects joints 1 and 3 of a frame and has no hinges, then the contribution of this member to the overall stiffness matrix is added into the locations shown in Figure 4.11 as K_{11}, K_{12}, K_{21} and K_{22}. If the member has a hinge at end A, then the contribution of the member to θ_1 column is repeated in the hinge column and the contribution of the member to M_1 row is also added to M_H row as shown in Figure 4.11. The reason for these operations concerning a hinge are given in detail in Chapter 6 when dealing with plastic hinges.

When all the contributions to the stiffness matrix have been included, the first $3m$ row and columns will normally each include contributions from two or more members. The other n row and columns, which correspond to hinge rotations, will each have received only the contribution from the member in which the hinge is located.

A flow diagram for the computer programme is shown in Figure 4.13. The non-linear analysis of a frame for several load factors

144

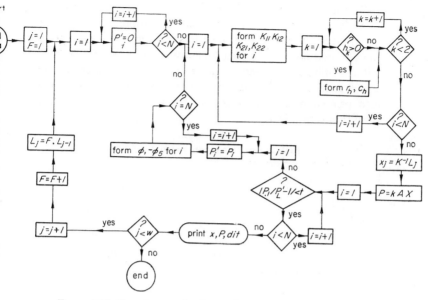

Figure 4.13. Flow diagram for the non-linear analysis of frames

starts by reading in the data for the problem. Under the initial set of applied loads the load factor F and the loading case j are both set to unity. At the first analysis of each loading case the axial loads P' are set to zero. This is followed by the construction of the overall stiffness matrix by adding the contribution of each member to the matrix. Each end k of a member is tested to see if it has a hinge and the hinge row r_h and hinge column c_h of the stiffness matrix for that hinge are also constructed. The equations $L = KX$ are then solved and the axial forces and bending moments are calculated for the members.

The value of the newly calculated axial force P_i in each member is compared to its value P'_i in the previous iteration and the iteration process is continued until for each member $|P/P'_i - 1|$ is less than the specified tolerance. If convergence is not obtained for each member the analysis is repeated under the same load factor. To do this the axial loads in the members are used to calculate the stability functions that are included in the overall stiffness matrix of the next analysis. For the purpose of calculating the stability functions the procedure given in Section 3.7 is utilised in the programme.

Whenever convergence is obtained, the values of the joint deflections, hinge rotations and axial loads are printed. The whole process is then repeated for a new load factor which is larger than
the old load factor by a fixed quantity such as 0·1. Every time the
stiffness matrix is solved its determinant is calculated and when
convergence is obtained the most recent value of the determinant is
also printed out. This value can be used to study the stability
condition of the frame.

Table 4.3 DATA FOR THE FRAME IN FIGURE 4.12

E	N	m	n	W			
207	3	2	1	1			
A	I	L	α	J_1	J_2	h_1	h_2
400	700	1700	− 60	2	0	0	7
200	500	1000	90	0	1	0	0
300	600	1000	30	1	2	0	0

Load vector			10	0	0	0	− 5	− 8		0
Tolerance			0·0001							

EXERCISES

1. Calculate the vertical deflection y_c of the propped cantilever
shown in Figure 4.14. Take $\lambda = 0·847, P = 121·56 \text{ N}, L = 508 \text{ mm}$,

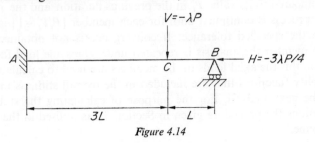

Figure 4.14

146

$E = 207$ kN/mm². The section is rectangular with breadth $b = 13$ mm and depth $d = 3$ mm.

Answer. $y_c/L = 0.126$

2. Calculate the sway deflection of the portal shown in Figure 4.15. The members are all 100 mm×100 mm square.

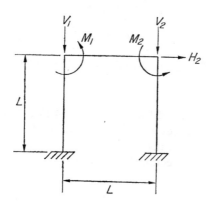

Figure 4.15

Take $\varrho = P/P_E = 0$ for the beam. $E = 207$ kN/mm²

$$V_1 = -50 \text{ kN}$$
$$M_1 = 21 \text{ kN.m}$$
$$V_2 = -100 \text{ kN}$$
$$M_2 = 29 \text{ kN.m}$$
$$H_2 = 5.2 \text{ kN}$$
$$L = 10 \text{ m}$$

Answer. $x_1 \simeq x_2 = 10$ mm

3. Calculate the joint rotation θ_B for the triangulated rigidly-jointed frame in Figure 4.16.

Answer. $\theta_B = 3(1+\sqrt{2})(\sqrt{2}\phi_{2,1}+\phi_{2,2})H/(\sqrt{2}\phi_{2,1}+\phi_{3,2})AE$

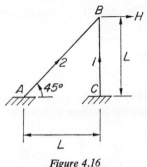

Figure 4.16

The second suffix in the stability functions refer to the member number.

4. The rigidly jointed frame in Figure 4.17 is simply supported at B and C and carries a vertical load W at A. If a moment M_A is applied at the apex calculate the joint rotations. All the members have the same cross-sectional properties.

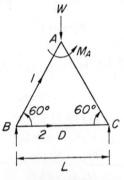

Figure 4.17

Answer. $\quad \theta_A = 0{\cdot}25L(2\phi_{3,\,1}+3\phi_{2,2})M_A/(\phi_{3,\,1}-\phi_{4.1}^2)EI$

$$\theta_B = \theta_C = -0{\cdot}25\phi_{4,\,1}LM_A/(\phi_{3,1}-\phi_{4,1}^2)EI$$

5. The rigidly jointed frame of Figure 4.18 is subject to a vertical force P acting upwards at A and forces $P/2$ acting downwards at B and C so that the members AB and AC are in tension and member BC is in compression.

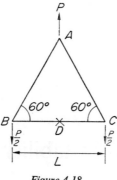

Figure 4.18

Calculate the resulting secondary moments at *A*, *B*, *C* and *D*, when
$P = 44\cdot480$ kN. The properties of the members are:

Property	AC & BC	AB
A	($322\cdot6$ mm^2)	($258\cdot1$ mm^2)
L	(539 mm)	(762 mm)
I	(4207 mm^4)	(4620 mm^4)
EI/L	($1\cdot588\times10^{-3}$ kNm)	($1\cdot255\times10^{-3}$ kNm)

Answer. $M_{BA} = M_{BC} = -M_{AB} = (1\cdot728\times10^3\,\text{kNm})$

$\qquad M_{CB} = -M_{CA} = (6\cdot068\times10^{-3}\,\text{kNm})$

$\qquad M_D = (6\cdot599\times10^{-3}\,\text{kNm})$

CHAPTER 5

Elastic Instability

Instability can be defined as the state where a structure loses its stiffness. Elastic instability takes place provided the material of the structure remains elastic. As was briefly described in Section 1.4, this may happen for idealised structures in one of two ways depending on the manner in which the external loads are applied. If these loads are applied axially to the members, then the structure does not deform until the critical load factor is reached, at which point undefined deformation may take place. On the other hand if the loads are applied transversely then the load deformation curve of the structure is non-linear culminating in the elastic failure load factor λ_{EF} when again the structure loses its stiffness.

In this chapter the state of elastic instability in both these two forms is studied, beginning again with the instability of individual members and then that of complete frames.

5.1. ELASTIC INSTABILITY OF MEMBERS

For a member in a general state of sway it was found in Section 4.1, equation (4.7) that the bending moment at end A of a member AB is given by

$$M_{AB} = 4k\phi_3\theta_A + 2k\phi_4\theta_B - \frac{6kv\phi_2}{L} \qquad (5.1)$$

where $k = EI/L$. The various modes of deformation of a member are special cases of this general state of sway. For this reason equation (5.1) will be used to derive the state of elastic instability with various specific modes of deformation.

5.1.1. Member with single curvature

In Figure 5.1(a), the member AB is subject to an axial load p and is deformed in single curvature due to the application of a moment M_{AB}. The rotation of end B, θ_{BA} is therfore opposite to that at A as

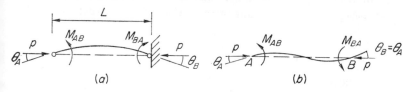

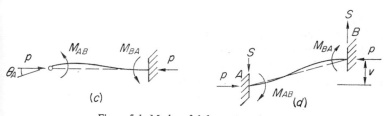

Figure 5.1. Modes of deformation of a member
(a) Pin-ended member in single curvature
(b) Member in double curvature
(c) Member pinned at one end, fixed at the other
(d) Member in pure sway

shown in the figure. Because there is no sway in the member, equation (5.1) for this case reduces to

$$M_{AB} = 4k\phi_3\theta_A - 2k\phi_4\theta_A \qquad (5.2)$$

An equation similar to (5.2) can be derived in terms of s and c functions either by using equations (3.7) and (3.11) or directly from (5.2) using the relationships of section (3.4).

The ratio $M_{AB}/\theta_A = 2(2\phi_3 - \phi_4)\,k$ is the stiffness of the member, and is defined as the moment required at end A to produce unit rotation at that end. As the axial load p is increased, this stiffness is dissipated and the condition of elastic instability is reached when

151

the stiffness vanishes, giving

$$\frac{M_{AB}}{\theta_A} = 2(2\phi_3 - \phi_4)k = 0 \tag{5.3}$$

The value of the axial load that brings about the elastic instability condition is called the elastic critical load P_c and under this load, equation (5.3) suggests that zero moment is required at A to rotate that end.

In Section 3.4 it is stated that $\phi_3 = (3\phi_2 + \phi_1)/4$ and $\phi_4 = (3\phi_2 - \phi_1)/2$. Substituting these in equation (5.3) gives:

$$2k\phi_1 = 0 \tag{5.4}$$

Thus the member becomes unstable when $\phi_1 = 0$ at $\varrho_c = P_c/P_E = 1$, that is to say at $P_c = P_E = \pi^2 EI/L^2$ which, as we know, is the Euler critical load of the member.

Denoting by K the stiffness of the member at any given instant and by K_0 the stiffness of the same member but without any axial load, i.e. when $\varrho = 0$ and $\phi_1 = 1$, then $K/K_0 = \phi_1$. A graph of K/K_0 against ϱ cuts the axis of ϱ when the latter is unity. This graph is the same as ϕ_1 against ϱ which is already shown in Figure 3.3.

5.1.2. Member in double curvature

When the same member is in double curvature as shown in Figure 5.1(b) under an axial load p and end moments M_{AB} and M_{BA} with $\theta_A = \theta_B$, equation (5.1) gives the stiffness of the member as

$$K = \frac{M_{AB}}{\theta_B} = 2(2\phi_3 + \phi_4)k \tag{5.5}$$

Substituting for ϕ_3 and ϕ_4, the state of instability, therefore results when

$$K = 2(2\phi_3 + \phi_4)k = 6k\phi_2 = 0 \tag{5.6}$$

This is satisfied when $\phi_2 = 0$ at $\varrho = 4$. Once again a graph of $K/K_0(= \phi_2)$ against ϱ is shown in Figure 3.3.

152

5.1.3. Member pinned at one end and fixed at the other

This is the case of the propped cantilever of Figure 5.1(c) where $\theta_B = v = 0$ and equation (5.1) reduces to

$$K = \frac{M_{AB}}{\theta_A} = 4k\phi_3 \tag{5.7}$$

and instability occurs when

$$K = 4k\phi_3 = 0 \tag{5.8}$$

This condition occurs when $\phi_3 = 0$ at $\varrho = 2 \cdot 046$. A graph of $K/K_0 = \phi_3$ against ϱ is also shown in Figure 3.3.

It is interesting to note that the bending moment at the fixed end of this propped cantilever is given by the last of equations (4.7) to be

$$M_{BA} = 2k\phi_4\theta_A \tag{5.9}$$

and the ratio M_{BA}/θ_A reduces to zero when ϕ_4 takes zero value. In Figure 3.3 it can be seen that ϕ_4 never becomes zero, indicating that such a state has no physical significance.

5.1.4. Member in pure sway

The translation stiffness of a member against sway is obtained by inducing pure sway by means of a pair of shear forces S as shown in Figure 5.1(d). From the second of equations (4.7), with $\theta_A = \theta_B = 0$, the shear force is given by

$$S = \frac{12k}{L^2} v\phi_5 \tag{5.10}$$

The quantity $K = S/v$ is the stiffness of the member with respect to lateral translation. A state of elastic instability in pure sway takes place when $S/v = 0$, thus:

$$K = \frac{12k}{L^2} \phi_5 = 0 \tag{5.11}$$

Now since $\phi_5 = \phi_1\phi_2$, it follows that ϕ_5, and therefore K, vanishes as soon as either ϕ_1 or ϕ_2 becomes zero. It is noticed in Figure 3.3 that ϕ_1 becomes zero at $\varrho = 1$ much before ϕ_2. Hence the elastic instability of the member takes place when $\phi_5 = \phi_1 = 0$. A graph

153

of $K/K_0 = \phi_5$ against ϱ is shown in Figure 3.3. This intersects the graph for ϕ_1 at $\varrho = 1$.

It is worth noting that each stability ϕ function may be considered as a measure of the stiffness of a member with a particular mode of deformation. Whenever one of these functions takes the value of zero, the member becomes unstable with a given mode. It is also observed that in every mode the instability of a member occurs when the forces required to produce that mode become zero. That is to say the stiffness of the member is lost and no external force is required to cause the said mode of deformation.

5.2. ELASTIC INSTABILITY OF FRAMES

The state of instability of a structure is similar to that of a member. A structure also becomes unstable when it loses all its stiffness and no external load is required to induce any desired mode of deformation in the structure. At such a stage, therefore, the externally applied load vector L becomes a null vector and the joint equilibrium $KX = L$ become:

$$KX = 0 \qquad (5.12)$$

These are a set of homogeneous linear equations whose solution is either given by

$$X = 0 = \{x_1 \; x_2 \; x_3 \; \ldots\},$$

or for 'non-trivial' values of $X \neq 0$, the determinant $|K|$ must vanish. Thus for the condition of instability of the structure we have

$$\text{Dit} = |K| = 0 \qquad (5.13)$$

This indicates that for an unstable structure, the set of joint equilibrium equations are no longer linearly independent.

For a structure that has n degrees of freedom the overall stiffness matrix K is of the order $n \times n$ and there are n equations in $L = K \cdot X$ to express the n different modes of deformation. It is possible to express[4] each linear elastic deformation x_o of a structure in the form of a series in terms of the first and higher critical modes x_{c1}, x_{c2} etc. thus

$$x_{0,\,1} = a_{11}x_{c1} + a_{12}x_{c2} + a_{13}x_{c3} \; \ldots \; a_{1n}x_{cn}$$

i.e. in matrix form

$$x_{0,\,1} = [a_{11} \; a_{12} \; a_{13} \; \ldots \; a_{1n}] \{x_{c1} \; x_{c2} \; \ldots \; x_{cn}\},$$

154

and the linear elastic column vector $X_0 = \{x_{0.1} \; x_{0,2} \ldots x_{0,n}\}$ is then given by

$$X_0 = aX_c \tag{5.14}$$

where a is an $n \times n$ matrix of constants and $X_c = \{x_{c1} \; x_{c2} \ldots x_{cn}\}$. In the case of non-linear analysis, the joint displacement vector X under a given load factor λ may be obtained by applying amplification factors v^{-1}, to the linear elastic displacements, thus

$$X = av^{-1}X_c \tag{5.15}$$

In equations (5.15) the matrix v is a diagonal $n \times n$ matrix with a typical element

$$v_i = \frac{\lambda_{ci} - \lambda}{\lambda_{ci}} \tag{5.16}$$

where λ_{ci} is the load factor for the ith instability condition in the structure.

Suppose n disturbing forces l are applied to a structure of n degrees of freedom such that $l_1 = vk_1x_1$, $l_2 = vk_2x_2$ etc. where k_1, k_2 etc. are constants of no fixed values, while v is a variable whose particular value under a given λ is obtained from (5.16). The set of equations $L = K.X$, with the structure being subject to the disturbing forces l become: $L = vkX = KX$.

Thus

$$[K - v\,k]X = 0 \tag{5.17}$$

where vk is a diagonal matrix with elements vk_1, vk_2 etc. The solution of equations (5.17) is either $X = 0$, or for non-trivial values.

$$\text{Dit} = |K - v\,k| = 0$$

This equation holds true irrespective of the values of the diagonal matrix k and therefore it is possible to replace k by a unit matrix.

Hence

$$\text{Dit} = |K - v\,I| = 0 \tag{5.18}$$

Equation (5.18) leads to an n degrees polynomial in v called the characteristic equation of the stiffness matrix K. The n different roots v_1, v_2 etc. of this equation are called the latent roots of the stiffness matrix and it is evident that whenever one of these roots becomes zero, equation (5.18) reduces to that given by (5.13). Thus one way of finding a condition of instability, for a structure, is to increase the axial loads in the members proportionally until one of the latent roots of the stiffness matrix becomes zero. The load factor at which this takes place is, by virtue of equation (5.16), the critical load factor λ_c. Equations (5.17) on the other hand give the critical vector X_c for the mode of deformation.

The preceding analysis contributes to the understanding of the characteristics of the stiffness matrix and its relationship to the critical load factors of the structure. In particular it shows that whenever a structure is unstable one of the latent roots of the stiffness matrix vanishes. There are various methods for determing the latent roots of a matrix, notably that by Jennings[5], which gives these roots simultaneously by an iterative procedure.

A particular load factor for instability can be determined by exciting a given mode of deformation in the structure by means of a single disturbing force and then using equations (5.13) to obtain the load factor at which the determinant $|K|$ becomes equal to zero. On the other hand when a structure is subject to a system of loads, the displacements in the structure are the result of all the modes of deformation as expressed by equations (5.15). As the load factor is increased, the non-linear load deflection curve sums up the influence of the entire load system in exciting all the various modes. The load factor reaches its maximum λ_{EF} at the elastic failure load and the condition of elastic instability is obtained again when $|K| = 0$. The details of the process for finding the condition of instability is now given.

5.3. DETERMINATION OF THE INSTABILITY CONDITION

The aim here is to obtain the value of the load factor for which the determinant of the stiffness matrix vanishes. To achieve this aim, the matrix displacement method for non-linear analysis given in section 4.2 can be used. At some convenient load factor λ the displacements in the structure are calculated taking the effect of the axial loads into consideration. These displacements are then substituted in $P = k \cdot A \cdot X$ to obtain a new set of axial loads, which if satisfactory are employed to calculate the stability functions of the members. These are then used to construct the overall stiffness matrix K of the structure and the determinant of this matrix is calculated. This procedure is followed with a number of load factors and a graph of the load factor against the values of the determinant is produced. The state of instability is obtained where this graph intersects the axis for the load factor. The operation can be speeded up by calculating the determinant at two arbitrary load factors and by a linear extrapolation the critical load factor is estimated. Further points on the graph are then obtained with load factors nearer to this estimated value. This can be done provided that care is taken that the load factors used are all relevant to the same state of instability.

The steps required by an iterative process are as follows:

1. Carry out a non-linear matrix displacement analysis going through the steps 1 to 6 of Section 4.2 with a selected value of the load factor $\lambda_i = \lambda_1$.

2. Construct the overall stiffness matrix and calculate its determinant D_i.

3. Repeat steps 1 and 2 with another value of the load factor λ_{i+1}

4. Interpolate or extrapolate linearly for a new value of the load factor λ_{i+2} at which the determinant of the stiffness matrix becomes zero. i.e. calculate λ_{i+2} from

$$\lambda_{i+2} = (D_i\lambda_{i+1} - \lambda_iD_{i+1})/(D_i - D_{i+1})$$

5. Replace $i+2$ for $i+1$ and $i+1$ for i and repeat steps 3 and 4 until the current value of λ is so near λ_c that step 6 of Section 4.2 does not converge after a number of solutions. It will be noticed at this stage that λ_{i+1} and λ_i are so near the load factor λ_{i+2} for instability that the interpolation of step 4 gives an acceptable value for λ_{i+2}. It is advantageous to have some values of λ for which the determinant of the stiffness matrix is negative and thus interpolate for the load factor at which instability takes place. The computer program of Section 4.6 prints out the value of the determinant of the stiffness matrix every time convergence takes place. Most computers are provided with facilities for calculating the determinant of the matrix of coefficients, for a set of simultaneous equations. It is therefore easy to modify the programme of Section 4.6 to obtain the state of instability.

157

5.4. EXTRACTION OF THE PARTICULAR MODE FROM THE GENERAL

In Section 5.2 it was pointed out that a particular load factor for instability with a given mode can be obtained by exciting the structure with a single disturbing force. In this case, instead of dealing with the entire overall stiffness matrix, it is possible to select only that part of the matrix which is relevant to the mode that is being considered. This makes the process of determining the condition of instability much easier as the rows and columns corresponding to the other modes are first eliminated. This process is best illustrated with the aid of an example.

Consider the portal frame of Figure 5.2 for this purpose. The equilibrium equations $L = K.X$ for this frame are

$$
\begin{bmatrix} H_1 \\ V_1 \\ M_1 \\ H_2 \\ V_2 \\ M_2 \end{bmatrix} = \begin{bmatrix} b_1+a_2 & & & \text{Symmetrical} & & \\ 0 & a_1+b_2 & & & & \\ -d_1 & -d_2 & e_1+e_2 & & & \\ -a_2 & 0 & 0 & a_2+b_3 & & \\ 0 & -b_2 & d_2 & 0 & b_2+a_3 & \\ 0 & -d_2 & f_2 & -d_3 & d_2 & e_2+e_3 \end{bmatrix} \begin{bmatrix} x_1 \\ y_1 \\ \theta_1 \\ x_2 \\ y_2 \\ \theta_2 \end{bmatrix}
$$

$$(5.19)$$

Assuming that the axial deformations of the members are negligible, the vertical movements y_1 and y_2 will therefore be small and the columns and rows corresponding to them can be excluded from equations (5.19). Resulting in

$$
\begin{bmatrix} H_1 \\ M_1 \\ H_2 \\ M_2 \end{bmatrix} = \begin{bmatrix} b_1+a_2 & & \text{Symmetrical} & \\ -d_1 & e_1+e_2 & & \\ -a_2 & 0 & a_2+b_3 & \\ 0 & f_2 & -d_3 & e_2+e_3 \end{bmatrix} \begin{bmatrix} x_1 \\ \theta_1 \\ x_2 \\ \theta_2 \end{bmatrix} \quad (5.20)
$$

To begin with if the side sway mode of deformation is suppressed, the rows and columns corresponding to x_1 and x_2 can be excluded from equations (5.20) leaving

$$
\begin{bmatrix} M_1 \\ M_2 \end{bmatrix} = \begin{bmatrix} e_1+e_2 & f_2 \\ f_2 & e_2+e_3 \end{bmatrix} \begin{bmatrix} \theta_1 \\ \theta_2 \end{bmatrix} \quad (5.21)
$$

158

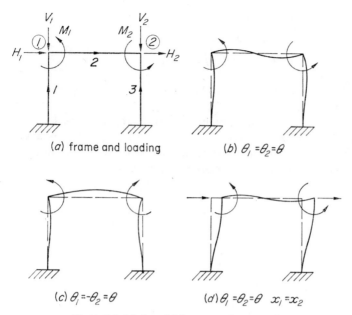

(a) frame and loading

(b) $\theta_1 = \theta_2 = \theta$

(c) $\theta_1 = -\theta_2 = \theta$

(d) $\theta_1 = \theta_2 = \theta$ $x_1 = x_2$

Figure 5.2. Modes of deformation for a portal

For a symmetrical frame, $e_1 = e_3$ and the antisymmetrical mode, of deformation is obtained when $\theta_1 = \theta_2$, a shown in Figure 5.2(b) and therefore only one equation in (5.21) is required to define this mode of deformation. The first one for instance, gives; for $M_1 = M_2 = M$, $\theta_1 = \theta_2 = \theta$

$$M = (e_1 + e_2 + f_2)\theta \qquad (5.22)$$

and the stiffness $K = M/\theta$ becomes zero when

$$e_1 + e_2 + f_2 = 0 \qquad (5.23)$$

i.e.

$$\frac{4E_1 I_1}{L_1}\phi_{3,1} + \frac{4E_2 I_2}{L_2}\phi_{3,2} + \frac{2E_2 I_2}{L_2}\phi_{4,2} = 0 \qquad (5.24)$$

where the second suffix in the stability functions refers to the member number.

The symmetrical mode for the same frame, while sway is still suppressed, is given by $\theta_1 = -\theta_2 = \theta$ as shown in Figure 5.2(c). Again one of the equations of (5.21) is sufficient to define this mode, giving

$$M = (e_1 + e_2 - f_2)\theta \qquad (5.25)$$

159

and the stiffness $K = M/\theta$ becomes zero when

$$\frac{4E_1I_1}{L_1}\,\phi_{3,1}+\frac{4E_2I_2}{L_2}\,\phi_{3,2}-\frac{2E_2I_2}{L_2}\,\phi_{4,2} = 0 \qquad (5.26)$$

On the other hand when the sway is not suppressed an antisymmetric mode of sway is obtained when $x_1 = x_2 = x$, and $\theta_1 = \theta_2 = \theta$, as shown in Figure 5.2(d). In this case the last two of equations (5.20) do not give us any new information about the frame that is not expressed by the first two. From these we obtain; remembering that the frame is itself symmetrical

$$\begin{bmatrix} H \\ M \end{bmatrix} = \begin{bmatrix} b_1 & -d_1 \\ -d_1 & e_1+e_2+f_2 \end{bmatrix} \begin{bmatrix} x \\ \theta \end{bmatrix}$$

The state of instability in this case is obtained when $H = M = 0$ i.e.

$$\mathrm{Dit} = \begin{vmatrix} b_1 & -d_1 \\ -d_1 & e_1+e_2+f_2 \end{vmatrix} = 0$$

giving

$$b_1(e_1+e_2+f_2)-d_1^2 = 0$$

substituting for b, e and f:

$$\frac{12E_1I_1}{L_1^3}\,\phi_{5,1}\left(\frac{4E_1I_1}{L_1}\,\phi_{3,1}+\frac{4E_2I_2}{L_2}\,\phi_{3,2}+\frac{2E_2I_2}{L_2}\,\phi_{4,2}\right)-\left(\frac{6E_1I_1}{L_1^2}\,\phi_{2,1}\right)^2$$
$$= 0$$

which is the condition for antisymmetric sway instability.

Once again the partitioning technique of Section 4.4 can be used here with advantage. Adopting the same notations of equations (4.14) and (4.15), the elements in the joint displacement vector concerning a given mode of deformation are first grouped together to correspond to X_2 of equations (4.14), rearranging the rows and the columns of the stiffness matrix accordingly. The condition of instability is then obtained by equating to zero the determinant of the resulting a_{22} submatrix. Once again the redundant rows and columns, due to the equality of some elements of X_2 are removed from a_{22}.

For the triangular frame of Figure 4.8 for example, the condition of instability for a pure rotation of joint B is given, from equations (4.13), by

$$|a_{22}| = e_1+e_2 = 0$$

160

i.e.

$$\frac{4E_1I_1}{L_1}\phi_{3,1} + \frac{4E_2I_2}{L_2}\phi_{3,2} = 0 \qquad (5.27)$$

For $\alpha = 30°$, $E_1I_1 = E_2I_2$, $L_1 = \sqrt{3}\,L_2 = L$, equation (5.27) gives:

$$\phi_{3,1} + \sqrt{3}\phi_{3,2} = 0 \qquad (5.28)$$

which gives the condition of instability.

5.5. APPROXIMATION OF THE AXIAL LOADS

As shown in Section 4.1, the axial loads in the members of a structure are unknown to begin with. The use of an iterative process in the analysis of the structure does, of course, overcome this difficulty. However, some knowledge of the axial loads not only simplifies the process of iteration but also saves computer time. Various suggestions have been made in this respect depending upon the type of structure.

For triangulated frames that are rendered statically determinate by inserting pins at their joints, it is sufficiently accurate to calculate the axial loads in the members from the pin-jointed structure as was suggested previously in Section 4.4. Thus, for the triangular frame of Figure 4.8, with $\alpha = 30°$, it is readily found that $\varrho_2 = \varrho_1/\sqrt{3}$. For various values of ϱ_1 the left-hand side of equation (5.28) was calculated and the results are given in Table 5.1. These figures were used in Figure 5.3 to plot a graph of ϱ_1 against $\phi_{3,1} + \sqrt{3}\phi_{3,2}$, from which the value of $\varrho_1 = 2\cdot63$ gives the condition for instability.

Table 5.1

ϱ_1	1	2	2·62	3
$\phi_{3,1} + \sqrt{3}\phi_{3,2}$	1·99	0·98	0·04	−0·87

For the rigidly jointed triangulated frame of Figure 4.9 and Section 4.4, the condition of instability is given by

$$|a_{22}| = \begin{vmatrix} e_3+e_4 & f_3 \\ f_3 & e_1+e_2+e_3 \end{vmatrix} = 0,$$

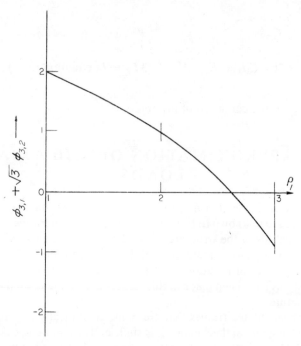

Figure 5.3. A graph of ϱ_1 againt $\phi_{3,1} + \sqrt{3}\phi_{3,2}$ for the frame of Figure 4.8

This gives

$$J = 4(\phi_{3,3}+\tfrac{1}{2}\sqrt{3}\phi_{3,4})\,(\sqrt{3}\phi_{3,1}+3\phi_{3,2}+\phi_{3,3})-\phi_{4,3}^2 = 0$$

It was shown in Section 4.4 that $\varrho_2 = \tfrac{2}{3}\,\varrho_1$, $\varrho_3 = 6\varrho_1$ and $\varrho_4 = -4\sqrt{3}$ ϱ_1 and for various values of ϱ_1, the values of J were calculated and a graph of J against ϱ_1 is shown in Figure 5.4. From this, the value of $\varrho_1 = 0{\cdot}43$ gives the condition of instability.

For rectangular rigidly-jointed frames it was considered by Merchant[6] that it is sufficiently accurate to neglect the axial forces in the beams and proportion the vertical loads between the columns so that the external columns equally carry half the load on the external bays above, while the internal columns equally share the rest. For example, for a single-storey frame with three bays carrying loads of W_1, W_2, and W_3 respectively, the external columns each carry $0{\cdot}25(W_1+W_3)$ while each internal column carries $0{\cdot}25(W_1+W_3)+0{\cdot}5W_2$. In order to justify this, it was suggested that for frames subject to vertical and horizontal loads, the movement of the vertical

162

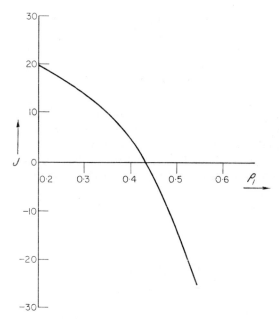

Figure 5.4. A graph of ϱ_1 against J for the frame shown in Figure 4.9

load from one side of the frame to the other strengthens some of the columns and weakens others with the net result that the response of the frame as a whole remains unaltered.

For instance, for the portal of Figure 5.2(a) with EI/L being equal for the three members, it will be assumed that $\varrho_1 = \varrho_3$ and $\varrho_2 = 0$. Therefore for the antisymmetrical mode Figure 5.2(b), equation (5.24) becomes

$$2\phi_{3,1}+3 = 0$$

Thus instability takes place at $\phi_{3,1} = -1\cdot5$ with $\varrho_1 = 3\cdot0953$.

Similarly, for the symmetrical mode of Figure 5.2(c), instability takes place at $\phi_{3,1} = -\frac{1}{2}$ while for the sway mode, Figure 5.2(d), instability occurs when

$$\phi_{5,1}(4\phi_{3,1}+6) = 3\phi_{2,1}^2$$

The reader can now calculate the values of ϱ_1 for these cases.

The approximation for the calculation of axial forces becomes unreliable in the case of rigidly-jointed frames with inclined members. For this type of frame, Horne[1] suggested estimating the distribution of axial loads in the members from a rigid-plastic analysis, for at the state of rigid plastic collapse, a mechanism develops in the frame rendering it statically determinate. This suggestion is reasonably satisfactory for the calculation of the elastic critical loads for use in the estimation of the failure loads of frames by a Rankine type approach. This is discussed later in Chapter 7.

It is obvious that, due to variations in the type of collapse mechanism and the position of the last hinge at collapse, some deviation will exist between the assumed and actual axial loads. Nevertheless it was noticed by the author[7] that for portals the elastic critical loads of practical frames agree to within 95% with the approximation suggested by Merchant. This assumption is therefore used in the next section to evaluate the critical loads for a family of pitched roof portal frames.

5.6. ELASTIC CRITICAL LOADS OF PITCHED ROOF PORTAL FRAMES

The pitched roof frame $ABCDE$ in Figure 5.5 is subject to axial forces shown in Figure 5.5(a). Two small disturbing forces Q excite a symmetrical mode of deformation in the frame and this is shown in Figure 5.5(b). On the other hand a pair of horizontal forces Q^1 acting as shown in Figure 5.5(c) induce an antisymmetrical mode. In the case of the symmetrical mode, (Figure 5.5(d)) the vertical displacement y_2 of the apex, can be expressed in terms of the horizontal displacement x_1 at the eaves level as

$$y_2 = x_1 \cot \gamma \tag{5.29}$$

This follows directly from the displacements vector diagram. In this mode the joint rotation θ_2 and the horizontal displacement x_2 at the apex are both zero. Thus neglecting the axial deformations of the members, since these have no effect on the state of instability, the remaining joint displacement vector for half the frame, shown in Figure 5.5(d), becomes $X = \{x_1 \ \theta_1\}$. The overall stiffness matrix a_{22}, therefore becomes:

$$a_{22} = \begin{bmatrix} b_1 + b_2 \operatorname{cosec}^2 \gamma & -d_1 + d_2 \operatorname{cosec} \gamma \\ -d_1 + d_2 \operatorname{cosec} \gamma & e_1 + e_2 \end{bmatrix} \tag{5.30}$$

The condition of instability is given when the determinant of a_{22}

164

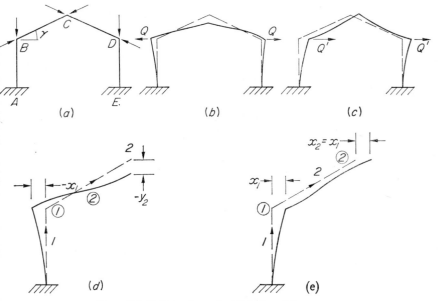

Figure 5.5. Deformations of a pitched roof frame
(a) Frame and loading
(b) Symmetrical mode
(c) Anti-symmetrical mode
(d) Numbering of members and joints and deformations
(e) Deformations for the anti-symmetrical mode

vanishes, hence:

$$(b_1 \sin^2 \gamma + b_2)(e_1 + e_2) - (d_2 - d_1 \sin \gamma)^2 = 0 \qquad (5.31)$$

The deformations and the numbering of the joints and members, for the antisymmetrical mode, are shown in Figure 5.5(e). For this mode the horizontal displacements at the eaves level x_1 and that at the apex x_2 are equal. Thus, once again neglecting the axial deformations of the members, the inclined member 2 will have no lateral sway with respect to its original inclination. In other words this member is in pure rotation and a rotation of θ_1 at joint 1 results in a rotation of $-c_2\theta_1$ at joint 2, i.e.

$$\theta_2 = -c_2\theta_1 \qquad (5.32)$$

Once again the joint displacement vector for this mode, considering half of the frame only, reduces to two elements $X = \{x_1 \; \theta_1\}$. The

165

overall stiffness matrix a_{22} for this mode therefore becomes

$$a_{22} = \begin{bmatrix} b_1 & -d_1 \\ -d_1 & e_1+e_2(1+c_2^2)-2c_2f_2 \end{bmatrix} \qquad (5.33)$$

From this, it can be shown that the condition of instability is given by

$$\phi_{1,1}\left[e_1+e_2-\frac{k_2\phi_{4,2}^2}{\phi_{3,2}}\right]-3k_1\phi_{2,1}=0 \qquad (5.33a)$$

where k_1 and k_2 are the EI/L values of members one and two respectively and the second suffix of the stability functions refers to the member number.

The axial loads in the members of a frame depend upon the applied load pattern. In Figure 5.6(a), a pitched roof frame is shown

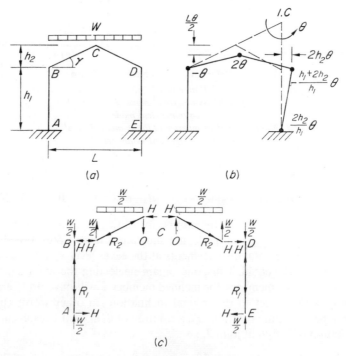

Figure 5.6. *Rigid plastic collapse mechanism of a pitched roof frame*
(a) Frame and loading
(b) Rigid plastic mechanism
(c) Member forces

166

subject to a uniformly distributed vertical load W. In order to calculate the axial loads in the members by the rigid plastic theory it is necessary to calculate the rigid plastic collapse load, W_p. In Figure 5.6(b) the collapse mechanism is shown with hinges at B, C, D and E. Giving the member CD an incremental deformation θ about the instantaneous centre I.C., the resulting rotations of the hinges would be as shown in the figure. From the geometry of the frame, the vertical movement of joint C downwards is readily calculated as $0 \cdot 5L\theta$ and the average movement of the load W would therefore be $0 \cdot 25L\theta$. The work done by this load is thus $0 \cdot 25L\theta$. The work absorbed by each hinge is given by the product of the hinge moment and its rotation. Thus the total work absorbed will be

$$M_P\left(\theta + 2\theta + \frac{h_1 + 2h_2}{h_1}\theta + \frac{2h_2}{h_1}\theta\right) = 4(1 + r)M_P\theta$$

where r is the ratio h_2/h_1. The virtual work equation would thus be

$$\tfrac{1}{4}LW\theta = 4(1 + r)M_P\theta$$

Hence
$$W_P = W = \frac{16(1 + r)M_P}{L} \tag{5.34}$$

The resulting internal member forces in the frame are shown in Figure 5.6(c). The average axial force R_2 in the rafters is calculated as follows. Considering member BC and taking moments about C, we have

$$WL/4 = Hh_2 + WL/8 + 2M_P$$

Hence using equation (5.34) and the fact that $h_2 = rh_1$ and $q = L/h_1$, the horizontal thrust H is given by

$$H = Wq/8(1 + r)$$

The axial force R_{2B} in member BC at end B of the member is therefore obtained by resolving the forces at B along BC. That is to say

$$R_{2B} = H\cos\gamma + 0 \cdot 5\, W\sin\gamma$$

Similarly the axial force R_{2C} in the same member but at end C is

$$R_{2C} = H\cos\gamma$$

Thus the average force R_2 in member BC is given by $R_2 = 0 \cdot 5(R_{2B} + R_{2C})$, i.e.

$$R_2 = H\cos\gamma + 0 \cdot 25\, W\sin\gamma$$

167

Hence using the value of H

$$R_2 = \frac{Wq \cos \gamma}{8(1+r)} + \frac{W \sin \gamma}{4}$$

and $$R_1 = \tfrac{1}{2}W \tag{5.35}$$

If the state of instability occurs when $W = W_c$, it follows that

$$\varrho_1 = \frac{R_1}{P_{E1}} = \frac{W_c}{2P_{E1}} = \frac{1}{2} W_c h_1^2 / \pi^2 EI \tag{5.36}$$

and $W_c = 2\varrho_1 P_{EI}$,
where P_{E1} and ϱ_1 refer to the columns, thus $P_{E1} = \pi^2 EI/h_1^2$. From the geometry of the frame the length BC of the rafters is $\tfrac{1}{2}qh_1/\cos \gamma$ and if ϱ_2 and P_{E2} refer to the rafters then

$$P_{E2} = \frac{4\pi^2 EI \cos^2 \gamma}{q^2 h_1^2}$$

i.e. $$\frac{P_{E1}}{P_{E2}} = \frac{1}{4} \frac{q^2}{\cos^2 \gamma} \tag{5.37}$$

Hence using equations (5.35) through (5.37) we obtain

$$\varrho_2 = \frac{q^2 \varrho_1}{8 \cos \gamma} \left[\frac{q}{2(1+r)} + \frac{2r}{q} \right] \tag{5.38}$$

For the same frame loaded with a point load W at the apex C, a similar procedure can be used to obtain values of ϱ in the members. These will be

$$\varrho_1 = \tfrac{1}{2}W_c h_1^2 / \pi^2 EI \tag{5.39}$$

and

$$\varrho_2 = \frac{q^2 \varrho_1}{4 \cos \gamma} \left[\frac{q}{2(1+r)} + \frac{2r}{q} \right] \tag{5.40}$$

Comparing equations (5.38) and (5.40) it is noticed that ϱ_2 for a frame loaded at the apex is twice that for a similar frame loaded uniformly. In both cases, however, the conditions of instability are given by equations (5.31) and (5.34).

Similar conditions can be established for the case of a pitched roof frame pinned at the supports. For instance it can be verified that elastic instability with an antisymmetrical mode occurs when

$$(1 - \phi_{1,1})(q\phi_{5,1}\phi_{3,2} + 2\phi_{3,1}\phi_{5,2} \cos \gamma) - \phi_{5,2}\phi_{1,1} \cos \gamma = 0 \tag{5.41}$$

and the condition for the case of symmetrical mode can also be found by a similar procedure. The axial forces in the case of a pinned base pitched roof frame can be obtained from

$$\varrho_1 = \frac{W_c h_1^2}{2\pi^2 EI},$$

$$\varrho_2 = \frac{q^2 \varrho_1}{8 \cos \gamma} \left[\frac{q}{2(2+r)} + \frac{2r}{q} \right] \tag{5.42}$$

In Figures 5.7 and 5.8 the value of ϱ that gives the condition of instability for various types of pitched roof frames are shown. From these graphs the value of ϱ at the elastic critical load, of a frame with a given ratio $q = L/h_1$, can be read off directly. Equation (5.36) then gives the elastic critical load. Figure 5.8 shows a comparison between values of ϱ_1 with a given pitch of $\gamma = 22\frac{1}{2}°$ but with different support and loading conditions. All the graphs show that the effect of axial load on the stability of the frames becomes more pronounced when the value of q increases. They also show that the state of instability with antisymmetrical mode takes place with a smaller external load than the symmetrical mode.

When calculating the condition of instability both in this section and in Section 5.5; once the distribution of the axial loads in the members of a structure has been decided, it is assumed that increasing the applied load factor does not alter this distribution. This means that the axial load in a member is linearly related to the external load factor. When using a computer, however, none of these assumptions are strictly necessary, since at any instant the computer programme calculates the exact values of the axial loads. However the assumptions are useful when carrying out an analysis manually. They are also useful when using a computer, since some computer time is saved by reducing the number of cycles of the iteration procedure.

As an example the pitched roof frame of Figure 4.4 is once again considered. The value of q for this frame is 1·848 and from the graph of Figure 5.8 the value of ϱ_1 for the symmetrical case at the critical load is 0·86. Hence the value of the vertical load $V_c = 2 \varrho_1 P_E$ is 1·60. kN.

Using the rigid plastic theory to calculate the axial loads in the members, equation (5.40) gives

$$\varrho_2/\varrho_1 = 1$$

169

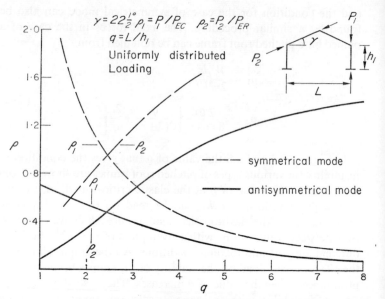

Figure 5.7. Elastic critical load of pitch roof frames

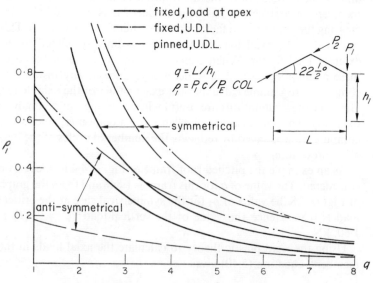

Figure 5.8. Elastic critical loads for pitched roof portal frames

An accurate non-linear analysis of this frame with various values of the external load, was carried out in Section 4.3 and the variation of the axial loads in the members with the external load is shown in Figure 4.5(b). It is noticed that the axial loads are nearly linearly related to the external load.

5.7. ELASTIC FAILURE LOAD OF TRUSSES

It was shown in Section 1.4 that flexible triangulated frames, with small primary bending moments in the members may remain elastic until collapse takes place at the elastic failure load λ_{EF}. A computer programme based on the iterative procedure of Sections 5.3 and 4.2 was used to calculate the elastic failure loads of a number of flexible trusses.

A series of tests on five truss-type frames carried out at Manchester University by Montague and Wardley[8], show close agreement with the results of the computer analysis. The dimensions and the manner of loading of the frames are shown in Figure 5.9. The members were manufactured out of small rectangular sections ranging between 3 mm × 3 mm to 3 mm × 13 mm sections and were welded to each other at the nodes by means of 20 mm diameter circular steel gusset plates.

The frames were analysed twice, first neglecting the gusset plate effects, and then including them and in Table 4.2 the results of the analyses are shown along with the experimental ones. It is noticed that neglecting the effects of the gusset plates reduces the calculated failure loads of the frames.

The results also show that including these effects renders the analytical results higher than those obtained experimentally. This slight difference is due to two reasons. Theoretically it is not so easy to calculate the stiffness of these small circular gussets and represent them accurately in the data. The high values of the analytical results may be therefore due to overestimation of the gusset plate stiffness. On the other hand it is not so easy to manufacture small frames of the types tested without introducing some small imperfections that invariably reduce the carrying capacity of structures by inducing more than one mode of deformations in the frame. Local yielding of some of the members particularly at high values of the load factor cannot be ruled out either.

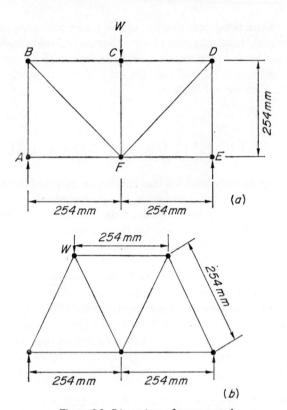

Figure 5.9. Dimensions of truses tested

(a) Pratt truss   (b) Warren truss

Table 5.2 EXPERIMENTAL AND THEORETICAL ELASTIC FAILURE LOADS OF TRUSSES
IN KN.

Frame No.	Experimental results	Theoretical results	
		Ignoring gussets	with gussets
1	1·33	1·30	1·42
2	2·17	2·04	2·20
3	2·16	2·03	2·18
4	1·67	1·59	1·79
5	1·65	1·58	1·76

Nevertheless it is conclusive that the small and stiff gusset plates
can increase the carrying capacity of these frames by as much as

7 per cent. When these gusset plates are taken into consideration the agreement between the theoretical and the experimental results is reasonable. The maximum difference being 7 per cent.

5.8. A QUICK DETERMINATION OF INSTABILITY CONDITION[7]

In this section it will be shown that it is possible to calculate the elastic critical load of a frame by considering only one value of the load factor λ.

In Section 5.2 the linear elastic joint displacement vector X_0 was expressed in equations (5.14) in terms of the first and higher critical modes X_c. Normally the first critical mode x_{c1} dominates the others and these can be neglected. Therefore, the joint displacement vector X_0 may be expressed in terms of the first critical mode, thus reducing equation (5.14) to

$$X_0 = x_{c1}\{a_{i1}\} \tag{5.43}$$

where once again $X_0 = \{x_{0,1}\, x_{0,2} \ldots x_{0,i} \ldots\}$, whereas $\{a_{i1}\} = \{a_{11}\, a_{21} \ldots a_{i1} \ldots\}$ is the first column of matrix a of equation (5.14).

Similarly the non-linear elastic displacement X may be obtained by applying a single amplification factor $1/v$ thus

$$X = \frac{x_{c1}}{v}\{a_{i1}\} \tag{5.44}$$

where in this case, from equation (5.16);

$$v = \frac{\lambda_{cl} - \lambda}{\lambda_{c1}} \tag{5.45}$$

As before λ is the applied load factor that causes X and λ_{c1} is the load factor for the first instability condition.

This information may be used for a quick determination of the condition at which instability takes place with the lowest load factor. The method however is not restricted to the lowest load factor and the procedure used can be adopted for higher load factors for instability. In order to explain the method the above mathematical expressions are first visualised graphically.

In Figure 5.10, the graph of a typical displacement δ obtained by a linear elastic analysis of a structure is shown against the load

173

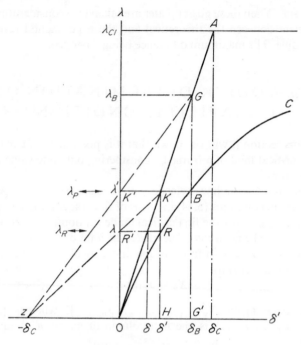

Figure 5.10. Graph of a typical displacement δ

factor λ. At the condition of instability at the load factor λ_{c1}, the value of δ becomes δ_c. For a given λ the linear relationship is given by

$$\delta = \delta_c \lambda / \lambda_{c1} \qquad (5.46)$$

Amplifying this displacement, using equation (5.45) gives

$$\delta^1 = \frac{\delta}{1 - \lambda / \lambda_{c1}} \qquad (5.47)$$

Substituting for δ from equation (5.46) into (5.47) gives

$$\delta^1 = \frac{\lambda \delta_c}{\lambda_{c1} - \lambda} \qquad (5.48)$$

where δ^1 is the displacement given by point R on the non-linear curve ORC under the load factor λ as shown in the figure.

Points such as R on the non-linear curve can be obtained graphically as follows. Point Z, which is at a distance of $-\delta_c$ from the origin, is joined to R^1 and line ZR^1 is extended to K on the linear elastic

174

line. A vertical line from K and a horizontal line from R^1 intersect at R on the non-linear elastic curve. This is because the resulting similar triangles KRR^1 and R^1OZ give

$$\delta^1 = \delta_c(\lambda^1 - \lambda)/\lambda \qquad (5.49)$$

Substituting $\delta^1\lambda_{c1}/\delta_c$ for λ^1, using the linear relationship (5.46), in equation (5.49), the same result as in equation (5.48) is obtained for δ^1. In this manner the non-linear elastic curve can be constructed with the aid of the linear elastic line. Thus in Figure 5.10, points such as K and G on the linear elastic line are used together with point Z in order to construct points such as R and B on the non-linear elastic curve.

It is evident that once point Z is located the non-linear load/displacement relationship of a structure can be readily obtained from its linear elastic load/displacement relationship. Hitherto, point Z has been obtained by first evaluating the elastic critical load. As was shown earlier in this chapter, this involves the extensive use of stability functions in an iterative procedure.

Figure 5.10, however, suggests that one iterative process to locate simultaneously any two points such as K and B at a constant load factor would result in the location of point Z and hence the elastic critical load of the structure would be evaluated directly. This is because once point K is located the linear elastic line $OKGA$ is obtained. From this line it is seen that

$$\lambda_B = \lambda^1\delta_B/\delta^1 \qquad (5.50)$$

Also from the similar triangles GG^1Z and K^1OZ

$$\lambda_B = \lambda^1(\delta_B + \delta_c)/\delta_c \qquad (5.51)$$

Equating the right-hand sides of equations (5.50) and (5.51) gives

$$\lambda^1\delta_B/\delta^1 = \lambda^1(\delta_B + \delta_c)/\delta_c$$

Hence

$$\delta_c = \frac{\delta^1 \times \delta_B}{\delta_B - \delta^1} \qquad (5.52)$$

But $\lambda_B/\delta_B = \lambda_{c1}/\delta_c$, therefore from equations (5.50) and (5.52) it follows that

$$\lambda_{c1} = \lambda^1\delta_c/\delta^1 = \frac{\lambda^1\delta_B}{\delta_B - \delta^1} \qquad (5.53)$$

Hence at a given load parameter λ^1, once δ^1 and δ_B are both obtained, equation (5.53) can be used to determine the elastic critical load parameter λ_{c1}.

175

As an example the elastic critical load of the two-storey frame of Figure 1.10 is calculated here making use of a computer for the solution of the equations $L = K \cdot X$. In the first solution, the axial loads in the members being ignored, the non-dimensional ratio of the sway at roof level to member length δ/L was 0·0426. After four cycles of iteration the value of $|(p_i/p_i^1) - 1|$ for each member was less than the adopted tolerance of 0·0001. The final axial load in member i is p_i while p_i^1 is the axial load in the same member at the previous cycle. At this stage the iteration process was terminated by the computer and the non-dimensional sway deformation at roof level δ_B/L was 0·0602. With these results equation (5.53) with λ^1 being unity gives

$$\lambda_{c1} = \frac{0·0602}{0·0602 - 0·0426} = 3·42$$

The same calculation carried out at the first floor level gave λ_c as 3.32. This difference, though small, is significant for two reasons. Firstly, it is due to the loading pattern exciting more than one mode of deformation in the frame and the components of these modes affect the frame differently at the two beam levels. Thus, to improve the results the loading pattern should be carefully selected to induce the same mode throughout the frame. For instance, to induce a sway deformation the vertical loads acting at the midspan of the beams should be moved to the top of the columns.

Secondly, as the value of the load factor approaches the critical load factor λ_{c1}, the corresponding mode of deformation becomes more dominating and hence a better value of the critical load is obtained by carrying out the analysis at a load level nearer λ_{c1} than the one used here.

The method given in this section becomes more useful when used to determine the elastic critical load of large structures where the computer time increases markedly. The method was used to evaluate the elastic critical load of a fixed base five bay thirty-storey frame. This frame was designed by Heyman[9] and in fact has the same dimensions, sectional properties and loading as the first thirty storeys of the sixty-storey frame shown in Figure 4.6. The difference in the calculated values of the elastic critical loads at various storey levels was not very pronounced. The value of the elastic critical load factor calculated from the sway displacements at the top storey level was 3·464 times the loads shown in Figure 4.6.

1. *ABC* is a continuous beam pinned at *A* and *C* and simply supported at *B*, where a moment *M* can be applied. For $AB = 2BC$ show that the stiffness M/θ_B is given by

$$2\pi^2\varrho_2[2/(1-n_1)+1/(1-n_2)],$$

where suffixes 1 and 2 refer to *AB* and *BC* respectively. Calculate the elastic critical load *P* of the beam.

Answer. $P = 14 \cdot 8EI/(AB)^2$

2. Calculate the elastic critical load of the column shown in Figure 3.13

Answer. $P = \pi^2EI/4L^2$

3. Obtain the value of *H* that makes the frame in Figure 4.16 elastically unstable.

Answer. $H = 3 \cdot 22\pi^2EI/L^2$

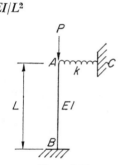

Figure 5.11

4. The column *AB* in Figure 5.11 is fixed at *B* and laterally supported at *A* by a spring which has an axial stiffness $k = \alpha EI/L^3$. Show that the condition of elastic instability is satisfied when

$$12\phi_5+\alpha = 9\phi_2^2/\phi_3.$$

Hence obtain the elastic critical load of a propped cantilever.

Answer. $P = 2 \cdot 045P_E$

5. The structure in Figure 5.12 consists of a fixed-ended continuous column *ABC* and a supporting pin-ended member *BD*. At *B* the structure is subject to a load *P* which is at 45° to the

column. Obtain the condition of elastic instability for the whole structure. For $L = 1$ metre; show whether elastic instability of the structure takes place before or after the pin ended member buckles on its own. Take $E = 207$ kN/mm². The two members are both $100^4 \sqrt{12}$mm$\times 100^4 \sqrt{12}$ mm square.

Answer. Condition of instability is given by

$$(b_1+b_2+EA/L)(e_1+e_2) = (d_1-d_2)^2$$

pin ended member buckles first when $P = 289000$ kN.

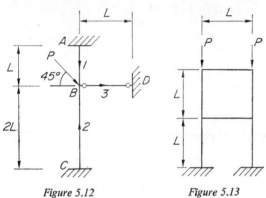

Figure 5.12   *Figure 5.13*

6. For the frame of 4.17 show that the two modes of elastic instability take place when

$$s_1+s_2+s_2c_2 = 0; \quad \text{for symmetrical mode};$$

$$s_1(s_1+s_2+s_2c_2)-(s_2c_2)^2 = 0, \quad \text{for antisymmetrical mode}.$$

The suffixes 1 and 2 refer to members AB and BC respectively.

7. Neglecting the axial loads in the beams of the two storey symmetrical frame in Figure 5.13 show that if side sway is prevented, the elastic instability conditions are given by

$$(1+2\phi_3)(1+4\phi_3) = 2\phi_4,$$

the members are identical.

8. Give the condition of elastic instability for the frame in Figure 5.13 when side sway is not prevented.

Answer.

$$(3\phi_2-2\phi_4\phi_1)^2 = (4\phi_3\phi_1-3\phi_2+6\phi_1)(8\phi_3\phi_1-6\phi_2+6\phi_1)$$

178

CHAPTER 6

Non-linearity due to Plasticity. Elastic-Plastic Analysis of frames

6.1. INTRODUCTION

In Chapter 1 it was shown that an important cause of non-linearity in the behaviour of structures is due to the development of material plasticity. The simplest way of dealing with the effect of plasticity is to assume that the material behaves in a rigid-plastic manner and then use the simple plastic theory to evaluate the collapse load of structures. The errors introduced in this approach were discussed in detail in Chapter 1 and it was suggested that a more realistic approach would be to assume that the material behaves in an elastic-plastic manner as shown by Figure 1.2(c). It was shown that, neglecting the axial load effect in a structure, the load displacement diagram of the structure may be idealised to that given by curve $OGHIJE$ of Figure 1.3 where the portions OG, GH etc. are linear.

The overall non-linearity of the load-deflection relationship, in this case, is purely due to the development of plastic hinges at discrete sections in the structure. Later on in Section 1.4 it was pointed out how the effect of axial loads in the members of a structure modifies the piece-wise linear elastic-plastic response in a non-linear manner, forcing the structure to collapse at a load factor below that given by the simple plastic theory and before the formation of sufficient hinges to convert the structure into a mechanism. This failure load was represented by point F on the elastic-plastic curve of Figure 1.9. Non-linearity due to the effect of axial loads was studied in detail in Chapters 3, 4 and 5 both for individual mbers and for complete structures.

In Section 3.5 the effects of axial load as well as the formation of plastic hinges in a member were also given where the slope deflection equations of a member were modified to include the effect of plastic hinges. In this chapter the effect of plastic hinge formation on the behaviour of structures as a whole is studied in detail and the elastic-plastic analysis of frames up to failure, taking the effect of axial loads into consideration, is presented.

It is useful to point out, at this stage, that one difference between an idealized plastic hinge and a real frictionless hinge is that, while a plastic hinge is capable of withstanding a constant bending moment equal to the plastic hinge moment of the section, a real hinge cannot withstand any moment. Thus, it can be assumed that a real hinge also behaves like a plastic hinge with its plastic hinge moment Mp being equal to zero. When a structure is subject to proportional loading, both its real and plastic hinges undergo continuous rotations. However, another difference between a plastic hinge and a real one is that, as the axial load in a member of a structure increases, a plastic hinge in that member withstands smaller moments and the plastic hinge moment of the section is reduced due to the effect of the axial load. The moment that a real hinge withstands remains unaltered and equal to zero irrespective of the value of the axial load in the member. If the axial load in a member is increased excessively it is possible to induce a plastic hinge to act as a real hinge by reducing the plastic hinge moment of the section to zero. The similarity between a real and a plastic hinge makes it possible to study the effect of both hinges on the behaviour of structures. This effect is studied in the next section.

6.2. THE DISPLACEMENT TRANSFORMATION MATRIX FOR A MEMBER WITH A HINGE

Consider first the simple frame ABC of Figure 6.1(a) which is fixed at A and C and rigidly jointed at B. Due to the external forces the joint displacement vector X for this frame is $X = = \{x_B \ y_B \ \theta_B\}$. Because the support at A is fixed, it has no displacement. Thus

$$\{x_A \ y_A \ \theta_A\} = \{0 \ 0 \ 0\} \qquad (6.1)$$

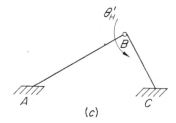

Figure 6.1. Hinge formation in a frame

(a) A rigidly jointed frame
(b) The frame with a hinge at end A of member AB
(c) The frame with a hinge at end B of member AB

If at some stage of loading a hinge is inserted at end A of the member AB, as shown in Figure 6.1(b), then this hinge begins to rotate as soon as the external loads are increased. The displacements of joint A are therefore altered to become

$$\{x_A \ y_A \ \theta_A\} = \{0 \ 0 \ \theta_H\}, \qquad (6.2)$$

where $\theta_A = \theta_H$ is the hinge rotation at support A. End A of

181

member AB is connected to the support at A and because of compatibility, total rotation θ_{AB} of this end of the member is given by

$$\theta_{AB} = 0 + \theta_H \qquad (6.3)$$

where the zero is the rotation of the fixed support before the insertion of the hinge.

Inserting the hinge gives the frame an extra degree of freedom θ_H and the joint displacement vector X of the whole frame becomes:

$$X = \{x_B \ y_B \ \theta_B \ \theta_H\} \qquad (6.4)$$

The external load vector for a structure is vectorially equivalent to the joint displacement vector and for each element of the joint displacement vector there is a corresponding element in the external load vector. For this frame the external load vector L that corresponds to the new displacement vector X, given by equation (6.4), is

$$L = \{H_B \ V_B \ M_B \ M_H\} \qquad (6.5)$$

The first three elements of this load vector are the external loads H_B, V_B and M_B applied at joint B as shown in Figure 6.1. The last element M_H is the moment that the hinge at end A of the member can withstand. In the case of a real hinge $M_H = 0$ while for a plastic hinge M_H is equal to the plastic hinge moment M_p of the section.

It should be pointed out at this stage that the development of a plastic hinge at a joint such as B, in Figure 6.1, has no meaning and joints have no plastic hinge moments. On the other hand each member of a frame meeting at a joint has a plastic hinge moment. Therefore when a frame is loaded, a plastic hinge may develop at the end of one of its members instead of the joint to which this end is connected.

If at a given stage of loading, a hinge is inserted at end B of member AB, as shown in Figure 6.1(c), then once the external loads are increased, the rotation θ_{BA} of end B of the member changes from $\theta_{BA} = \theta_B$ to become

$$\theta_{BA} = \theta_B + \theta_H^1, \qquad (6.6)$$

where θ_H^1 is the rotation of the hinge and θ_B is the rigid joint rotation. Once again inserting this hinge gives the frame an extra degree of freedom. This degree of freedom is the hinge rotation θ_H^1.

This convenient method of altering the compatibility equation for the end rotation of a member is applicable to any member of

182

a structure. Consider now a general member of a plane frame that is rigidly connected to joints i and j at its first and second ends respectively. The member displacement vector is

$$Z = \{u \ v \ \theta_{R1} \ \theta_{R2}\} \tag{6.7}$$

The displacement vector for the two joints is

$$X = \{x_i \ y_i \ \theta_i \ \ldots \ x_j \ y_j \ \theta_j\} \tag{6.8}$$

If there is no hinge at either end of the member, its displacement transformation matrix which relates the two sets of displacements is given by equation (2.34) of Section 2.3. Inserting a hinge at one end, say the second end, of the member does not change the member displacement vector (6.7) except that θ_{R2} now means

$$\theta_{R2} = \theta_j + \theta_{H2} \tag{6.9}$$

where θ_j is the joint rotation at j without a hinge and θ_{H2} is the hinge rotation.

However the joint displacement vector (6.8), does change. The extra degree of freedom θ_{H2} must be included in this vector which becomes

$$X = \{x_i \ y_i \ \theta_i \ \ldots \ x_j \ y_j \ \theta_j \ \theta_{H2}\} \tag{6.10}$$

The first three equations of (2.34) are unaltered but the last one is changed to become that given by equation (6.9) above. The new transformation matrix connecting the member and joint displacements, becomes

$$
\begin{bmatrix} u \\ v \\ \theta_{R1} \\ \theta_{R2} \end{bmatrix} =
\begin{bmatrix} -l_p & -m_P & 0 & \ldots & l_P & m_P & 0 & 0 \\ -l_Q & -m_Q & 0 & \ldots & l_Q & m_Q & 0 & 0 \\ 0 & 0 & 1 & \ldots & 0 & 0 & 0 & 0 \\ 0 & 0 & 0 & \ldots & 0 & 0 & 1 & 1 \end{bmatrix}
\begin{bmatrix} x_i \\ y_i \\ \theta_i \\ x_j \\ y_j \\ \theta_j \\ \theta_{H2} \end{bmatrix} \tag{6.11}
$$

It is noticed that the displacement transformation matrix A has a new column $\{0 \ 0 \ 0 \ 1\}$. The external load vector L corresponding to (6.10) is

$$L = \{H_i \ V_i \ M_i \ \ldots \ H_j \ V_j \ M_j \ M_H\} \tag{6.12}$$

where M_H is equal to zero for a real hinge or Mp of the member for a plastic hinge.

Similarly if the hinge is inserted at the first end of the member, the relationship $Z = AX$ becomes

$$
\begin{bmatrix} u \\[2mm] v \\[6mm] \theta_{R1} \\[2mm] \theta_{R2} \end{bmatrix} =
\begin{bmatrix}
-lp & -m_P & 0 & 0 & \ldots & lp & m_P & 0 \\[2mm]
-l_Q & -m_Q & 0 & 0 & \ldots & l_Q & m_Q & 0 \\[4mm]
0 & 0 & 1 & 1 & \ldots & 0 & 0 & 0 \\[2mm]
0 & 0 & 0 & 0 & \ldots & 0 & 0 & 1
\end{bmatrix}
\begin{bmatrix} x_i \\ y_i \\ \theta_i \\ \theta_{H1} \\ \vdots \\ x_j \\ y_j \\ \theta_j \end{bmatrix} \quad (6.13)
$$

The new column $\{0\ 0\ 1\ 0\}$ of the A matrix corresponds to the hinge rotation θ_{H1} at the first end of the member. The reader may now construct the displacement transformation matrix for a member with hinges at both ends.

Unlike the displacement transformation matrix it should be pointed out that the stiffness matrix k of the member does not change. This is natural enough as the stiffness of a member is not dependent on any hinges at its ends but on the member properties such as length, area, second moment of area and modulus of elasticity. The stiffness matrix of a member of a rigidly jointed frame, with or without hinges is, as before, given by equation (2.8) when the axial load is neglected and by equation (4.4) or (4.7) if the axial load is considered.

6.3. THE OVERALL STIFFNESS MATRIX

In the previous section it was pointed out that the member stiffness matrix does not alter because of inserting hinges at its ends. The displacement transformation matrix changes in the manner shown by equations (6.11) and (6.13) and because of this the overall stiffness matrix of the structure changes. The contribution of a member connected to joints i and j, and with a hinge at its second end, to the overall stiffness matrix can be derived by first transposing matrix A of equation (6.11) and then carrying out the triple matrix multiplication A^1kA in the usual manner. This

184

operation produces the contributions of the member to the joint equilibrium equation $L = KX$ as

$$
\begin{array}{l}
\text{at} \\
\text{joint} \\
i \\
\\
\\
\text{at} \\
\text{joint} \\
j \\
\\
\text{at hinge}
\end{array}
\begin{bmatrix}
H_i \\
V_i \\
M_i \\
\vdots \\
\vdots \\
H_j \\
V_j \\
M_j \\
\hdashline
M_H
\end{bmatrix}
$$

$$
=
\begin{bmatrix}
A & B & -C & \cdots & -A & -B & -C & -C \\
B & F & -T & \cdots & -B & -F & -T & -T \\
-C & -T & e & \cdots & C & T & f & f \\
& & & & & & & \\
-A & -B & C & \cdots & A & B & C & C \\
-B & -F & T & \cdots & B & F & T & T \\
-C & -T & f & \cdots & C & T & e & e \\
\hdashline
-C & -T & f & \cdots & C & T & e & e
\end{bmatrix}
\begin{bmatrix}
x_i \\
y_i \\
\theta_i \\
\vdots \\
x_j \\
y_j \\
\theta_j \\
\hdashline
\theta_{H2}
\end{bmatrix}
\quad (6.14)
$$

at joint i at joint j

Ignoring the effect of the axial load in the member, the quantities A, B, C, F, T, e and f are defined by equations (2.9) and (2.42). On the other hand when taking the effect of the axial load into consideration, these quantities are defined by equations (2.42) and (4.8).

It is interesting to compare the contribution of a member with a hinge to the overall stiffness matrix K with that of an elastic member without a hinge. An inspection of equations (6.14) and (2.41) shows at a glance that in equation (6.14) the large submatrix under joints i and j is exactly the same as that for an elastic member, as given by equation (2.41). It is also noticed that in equation (6.14) the coefficients of θ_j and θ_{H2} respectively given by the last two columns of the stiffness matrix are exactly the same. Thus inserting a hinge at the second end of a member causes the elements of the

185

member contributions, corresponding to θ_j in the overall stiffness matrix, to be repeated as elements corresponding to the hinge rotation θ_{H2}. Furthermore since the bending moment M_H at the hinge is equal to the bending moment at end j of the member it is noticed that the stiffness contributions of this member to the rows corresponding to M_j and M_H are also the same. In the case of a plastic hinge, M_H in the load vector of equations (6.14) is set equal to the plastic hinge moment of the section. For a real hinge M_H is set to zero.

Similarly in the case of a hinge at the first end of the member, the reader can verify that its contributions to the overall stiffness matrix are given by

$$
\begin{array}{l}
\text{at} \\
\text{joint} \\
i \\[4pt]
\text{at} \\
\text{joint} \\[4pt]
\text{at hinge}
\end{array}
\begin{bmatrix}
H_i \\
V_i \\
M_i \\
\cdot \\
\cdot \\
H_j \\
V_j \\
M_j \\
\hdashline
M_H
\end{bmatrix}
$$

$$
=
\begin{bmatrix}
A & B & -C & \cdots & -A & -B & -C & \vdots & -C \\
B & F & -T & \cdots & -B & -F & -T & \vdots & -T \\
-C & -T & e & \cdots & C & T & f & \vdots & e \\
 & & & & & & & \vdots & \cdot \\
 & & & & & & & & \cdot \\
-A & -B & C & \cdots & A & B & C & \vdots & C \\
-B & -F & T & \cdots & B & F & T & \vdots & T \\
-C & -T & f & \cdots & C & T & e & \vdots & f \\
\hdashline
-C & -T & e & \cdots & C & T & f & \vdots & e
\end{bmatrix}
\begin{bmatrix}
x_i \\
y_i \\
\theta_i \\
\cdot \\
\cdot \\
x_j \\
y_j \\
\theta_j \\
\hdashline
\theta_{H1}
\end{bmatrix}
\quad (6.15)
$$

where the spanning labels "at joint i" and "at joint j" head the columns.

where again it is noticed that the columns corresponding to θ_i and θ_{H1} are identical and the rows corresponding to M_i and M_H are also identical. The arrows in equations (6.14) and (6.15) indicate which rows and columns are repeated.

A layout of the overall stiffness matrix is shown in Figure 4.11. This was given in section 4.6 where a computer programme for the non-linear elastic analysis of plane frames with hinges was described. The matrix is of order $(3m+n)\times(3m+n)$, where m is the total number of joints in the frame and n is the number of real or plastic hinges. Accordingly the load vector L consists of $(3m+n)$ elements of which the first $3m$ elements are the external loads and the last n elements are the hinge moments M_{H1} to M_{Hn} for a total of n hinges in the frame.

6.4. ELASTIC-PLASTIC ANALYSIS OF A PROPPED CANTILEVER

As an introduction to the elastic-plastic analysis of frames the simple propped cantilever of Figure 6.2(a) is analysed as an example. The member can be treated as a structure with two members AC and CB with joint C being rigid and joint B pinned. The members and joints are numbered in the usual manner as shown in the figure. The joint displacement vector is

$$X = \{x_1 \ y_1 \ \theta_1 \ x_2 \ \theta_2\} \tag{6.16}$$

The horizontal movements x_1 and x_2 may, at this stage of the analysis, be excluded from this vector. If the horizontal force H_2 is not acting, the corresponding external load vector is:

$$L = \{H_1 \ V_1 \ M_1 \ H_2 \ M_2\} = \{0 \ -\lambda P \ 0 \ 0 \ 0\} \tag{6.17}$$

where λ is the load factor which takes the value of unity at the rigid plastic collapse load of the frame. The rigid plastic collapse mechanism is shown in Figure 6.2(b). The virtual work equation for an incremental rotation θ is

$$Mp\theta + Mp(\theta+3\theta) = 3\lambda PL\theta$$

Hence for $\lambda = 1$

$$Mp = 3PL/5 \tag{6.18}$$

The joint equilibrium equations $L = K \cdot X$ are constructed in the

187

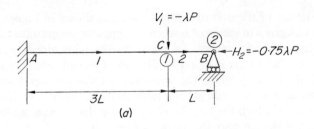

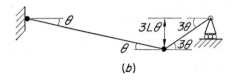

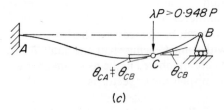

Figure 6.2. A propped cantilever
(a) Dimensions and numbering of members and joints
(b) Rigid-plastic collapse mechanism
(c) Deformed shape after the formation of a plastic hinge at C

usual manner. These are

$$
\begin{array}{c}
\text{at} \\
\text{joint} \\
1
\end{array}
\begin{array}{c}
\text{at} \\
\text{joint 2}
\end{array}
\begin{bmatrix}
0 \\
-\lambda P \\
0 \\
\hline
0 \\
0
\end{bmatrix}
=
\begin{bmatrix}
a_1+a_2 & 0 & 0 & -a_2 & 0 \\
0 & b_1+b_2 & d_1-d_2 & 0 & -d_2 \\
0 & d_1-d_2 & e_1+e_2 & 0 & f_2 \\
\hline
-a_2 & 0 & 0 & a_2 & 0 \\
0 & -d_2 & f_2 & 0 & e_2
\end{bmatrix}
\begin{bmatrix}
x_1 \\
y_1 \\
\theta_1 \\
\hline
x_2 \\
\theta_2
\end{bmatrix}
$$

(6.19)

where it is noticed that the first and fourth equations, corresponding to the horizontal movements x_1 and x_2, can be removed without affecting the other equations. Solving the latter as linear

188

equations gives

$$y_1 = -117\lambda PL^3/256EI, $$
$$\theta_1 = 63\lambda PL^2/256EI, \qquad\qquad (6.20)$$
$$\theta_2 = 144\lambda PL^2/256EI$$

The member forces can be calculated from $\boldsymbol{P} = \boldsymbol{kAX}$ which gives

$$M_{AC} = 60\lambda PL/128$$
$$M_{CA} = -M_{CB} = 81\lambda PL/128 \qquad\qquad (6.21)$$
$$M_{BC} = 0$$

As the load factor λ increases the elastic member reaches a stage where a plastic hinge develops at the point of maximum bending moment. From equations (6.21), this occurs at C under the vertical load and at this stage M_{CA} becomes equal to Mp. Thus using (6.18)

$$Mp = 3PL/5 = 81\lambda PL/128$$

giving:

$$\lambda = 0\cdot948 \qquad\qquad (6.22)$$

At this load factor the vertical elastic deflection y_{IE} under the applied load, from the first of equations (6.20), becomes

$$y_{IE} = -0\cdot433PL^3/EI \qquad\qquad (6.23)$$

As the load factor λ is increased beyond a value of $0\cdot948$ the plastic hinge at C begins to rotate under a constant moment of $M_C = M_P$. In Figure 6.2(c) the deformed shape of the structure is shown with the plastic hinge inserted at the second end C of member 1. It is indicated in the figure that continuity no longer exists at point C, thus the end rotation θ_{CA} is different from θ_{CB}. As the plastic hinge is inserted slightly to the left of joint 1, the rotation θ_{CB} is equal to the joint rotation θ_2. On the other hand the end rotation θ_{CA} of the second end of member 1 is equal to $\theta_1 + \theta_H$ where θ_H is the rotation of the plastic hinge. The new joint displacement vector, including θ_H as a new degree of freedom, is

$$X = \{x_1 \ y_1 \ \theta_1 \ x_2 \ \theta_2 \ \theta_H\} \qquad\qquad (6.24)$$

The corresponding external load vector is

$$L = \{H_1 \ V_1 \ M_1 \ H_2 \ M_2 \ M_H\} = \{0 \ -\lambda P \ 0 \ 0 \ 0 \ M_p\} \ (6.25)$$

It is noticed, in equations (6.21), that M_{CA} is positive and therefore $+M_p$ is inserted, in the load vector of equation (6.25). This is the plastic hinge moment of member 1.

When constructing the joint equilibrium equations $L = \boldsymbol{K} . \boldsymbol{X}$, the column elements corresponding to θ_H and the row elements corresponding to M_p are prepared in the manner given by equations (6.14), for a hinge at the second end of member 1. The joint equilibrium equations thus become

$$
\begin{array}{c}
\text{at} \\ \text{joint} \\ 1 \\ \hline \text{at} \\ \text{joint 2} \\ \hline \text{at hinge}
\end{array}
\begin{bmatrix}
0 \\ -\lambda P \\ 0 \\ \hline 0 \\ 0 \\ \hline M_p
\end{bmatrix}
$$

$$
=
\left[
\begin{array}{ccc|cc|c}
a_1+a_2 & 0 & 0 & -a_2 & 0 & 0 \\
0 & b_1+b_2 & d_1-d_2 & 0 & -d_2 & d_1 \\
0 & d_1-d_2 & e_1+e_2 & 0 & f_2 & e_1 \\
\hline
-a_2 & 0 & 0 & a_2 & 0 & 0 \\
0 & -d_2 & f_2 & 0 & e_2 & 0 \\
\hline
0 & d_1 & e_1 & 0 & 0 & e_1
\end{array}
\right]
\begin{bmatrix}
x_1 \\ y_1 \\ \theta_1 \\ \hline x_2 \\ \theta_2 \\ \hline \theta_H
\end{bmatrix}
\qquad (6.26)
$$

with headings *at joint 1*, *at joint 2*, *hinge* over the respective column groups.

Neglecting the effect of any axial load in the member and solving equations (6.26) for the vertical deflection y_1 under a load factor $\lambda = 0{\cdot}948$ gives $y_1 = 0{\cdot}433PL^3/EI$. This is the same value as that given by equation (6.23) obtained by the elastic analysis. However solving equations (6.26) with $\lambda = 1$ gives:

$$y_1 = -0{\cdot}9PL^3/EI \qquad (6.27)$$

This is the condition where a second plastic hinge develops at end A of the propped cantilever and a mechanism is developed at the rigid plastic load factor $\lambda = 1$. The elastic-plastic load-deflection diagram is shown non-dimensionally by the piece-wise linear curve $OCAE$ in Figure 6.3.

So far the axial load H_2, Figure 6.2(a), in the propped cantilever has been assumed to be zero. For non-zero values of this load, the behaviour of the structure is no longer piece-wise linear. A compressive axial load reduces the stiffness of the members and thus of the whole structure. Nevertheless the analysis can be carried out using equations (6.19), for the elastic stage, and (6.26), beyond

190

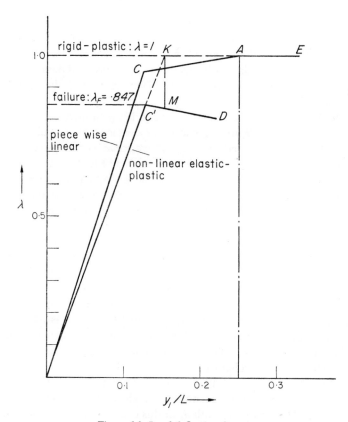

Figure 6.3. Load-deflection diagram

the formation of the first hinge, provided that the notations of equations (4.8) are adopted. These include the stability function of the members.

For the purpose of an elastic-plastic analysis with the axial loads being considered, the horizontal load H_2 is taken as $0\cdot75\ \lambda P$. Thus at any load factor

$$\varrho_1 = \frac{0\cdot75\lambda P}{\dfrac{\pi^2 EI}{9L^2}}$$

and

$$\varrho_2 = \frac{0\cdot75\lambda P}{\dfrac{\pi^2 EI}{L^2}}$$

(6.28)

191

That is to say $\varrho_1 = 9\varrho_2$. The total span $4L$ of the structure is taken as 508 mm, its modulus of elasticity as 207 kN/mm^2 and the full plastic moment as 9·27 kNmm, thus $P = 121\cdot56$ N. The section is taken to be rectangular with breadth b of 13 mm and depth d of 3 mm At the elastic stage equations (6.19) are used, with the stability functions for the members being calculated by using equations (6.28). Repeated solution of equations (6.19) with increasing values of λ shows that the first plastic hinge develops at a load factor $\lambda = 0\cdot847$. At this load the non-dimensional vertical deflection $|y_1/L|$ is 0·126. This corresponds to point C^1 in Figure 6.3.

Having inserted the plastic hinge at the second end of member 1, equations (6.26) are used for the analysis while the stability functions are still calculated using equations (6.28). It is found that the load factor cannot be increased above 0·847 and in fact the deflections increase by reducing the load factor as shown by curve C^1MD of Figure 6.3. Point C^1 therefore represents the failure state of the structure. It is significant, that, using the stability functions, the determinant of the overall stiffness matrix of equations (6.19) at $\lambda = 0\cdot847$ is positive while, having inserted the hinge, the determinant of the overall stiffness matrix of equations (6.26) is negative. This indicates that inserting the hinge at C renders the structure unstable and failure takes place with a single plastic hinge before the development of a mechanism.

Before leaving this example, two further complications should be pointed out concerning the non-linear elastic-plastic analysis of structures. The first point is that, in this example, the axial load in the member was known throughout the analysis at any load factor, since this axial load was applied externally at support B. In most structures the axial loads in the members are unknown and can only be found by an iterative process.

The second point is that these axial loads continuously reduce the plastic hinge moments of the member sections. In the example of the propped cantilever, the properties of the rectangular member were selected so that the full plastic moment of the section does not alter significantly. In structures where high axial loads exist and where I sections are used, the effect of these loads in reducing the plastic hinge moments of the sections is considerable. The axial load in a member thus speeds up the formation of a plastic hinge in it. After the formation of a hinge in a member, as the load factor increases, so does the axial load in the member and thus the plastic hinge moment of the section is reduced still further. For this reason, the use of equations (6.14) or (6.15) in the elastic-plastic analysis requires the substitution of the reduced plastic hinge moment M_p^1 of the members for M_H. The value of M_p^1 changes as the load factor is altered and a fresh calculation of its value is necessary at all stages of the analysis.

It is sometimes considered that the strain hardening effect of the material of the structure may compensate for some of the damaging effects of the axial loads. Strain hardening can be significant in practical structures having a few storeys. This is because, on the one hand, the axial loads in the members of these structures are small and hence their effects in reducing the stiffness of the structure and the plastic hinge moments of the sections are also small. On the other hand, these structures either collapse with the formation of a sufficient number of hinges for the development of a mechanism or the structure may collapse just before the development of such a mechanism. In both cases, there will be a number of plastic hinges with sufficient rotation for strain hardening to be significant. Taller structures with high axial loads collapse before the development of a mechanism and with only few hinges. The strain hardening effect in these hinges is not large enough to play any significant part.

6.5. ELASTIC-PLASTIC ANALYSIS OF FRAMES

Fundamentally the elastic-plastic analysis of frames in general is similar to the elastic-plastic analysis of the propped cantilever of the previous section. This consists of tracing the non-linear load-deflection history of the frame up to collapse. One way of proceeding with the elastic-plastic analysis of a frame is to vary the applied load factor by a regular increment until the maximum bending moment in the frame becomes equal to the plastic hinge moment of the section where it occurs. A plastic hinge is inserted at this section and the load factor is increased by the same regular increments until the bending moment at a second point reaches the plastic hinge moment of the section. This process is continued up to the stage where the frame loses all its stiffness at its elastic-plastic failure load. However, the defects and the difficulties facing such an approach are

1. The stiffness equations are solved too many times. Although a complete load-deflection curve is obtained, this curve is rarely used.

2. At high values of the load factor it is likely that more than one hinge will develop between two consecutive load factor. The insertion of these hinges all at once brings inaccuracies, and may change the mode of deformation fundamentally. To rectify this defect a smaller increment of the load factor may be adopted but this increases the computation time.

3. The procedure of regular load increments can only limit the collapse load between an upper and a lower bound, and the difference between these can be reduced only by more solutions of the stiffness matrix.

A more refined elastic-plastic analysis of a frame can be carried out by moving from one hinge to another. Once a plastic hinge is detected and inserted in the frame, the load factor at which the next plastic hinge would form is extrapolated and applied to the frame. The steps involved in such an analysis are

1. Initially the frame is assumed to be elastic and its overall stiffness matrix is constructed on the assumption that the axial forces in the members are zero. The contribution of one member to this matrix is given in equation (2.41) with the notations of equations (2.42) and (2.9).

2. The joint equilibrium equations $L = K \cdot X$ are solved with a given load factor λ_1.

3. The resulting joint displacements are used in equations (2.45) to calculate the bending moments and the axial forces in the members.

4. By a linear extrapolation between a load factor of zero and λ_1 the load factors at which the bending moment, at each end of every member, reaches the plastic hinge moment of the member is calculated. The lowest of these load factors is the expected factor λ_2 at which the next hinge is predicted to form in the frame.

5. Similarly, employing the axial loads calculated at step 3, the axial loads in the members under λ_2 are extrapolated.

6. These axial loads are used to calculate the stability functions for the members which are utilized to construct the overall stiffness matrix of the frame. The contribution of one member to this matrix is given by equations (2.41) and (2.42) but with the notations of equations (4.8).

194

7. The equations $L = K \cdot X$ are solved for the current load factor λ_2 and a new set of joint displacements are calculated.

8. Equations (2.45) are again employed to calculated the new axial loads and bending moments in the members.

9. The bending moments at either end of each member at the previous and current load factors λ_1 and λ_2 are used to interpolate or extrapolate the load factor at which a plastic hinge develops at that end of the member. Again the lowest of these load factors is the expected load factor λ at which the next hinge is predicted to develop in the frame.

10. From the axial loads in the members at load factors λ_1 and λ_2 the axial loads in the members at load factor λ are calculated.

11. The process is repeated from step 6. Each time the predicted load factor λ replaces the current load factor λ_2 and is used to solve $L = K \cdot X$. Thus λ_1 is the load factor at the previous cycle of iteration, λ_2 is the current applied load factor and λ is the predicted load factor for the next hinge. The stage will be reached when λ_2 and λ are within a specified small tolerance. This indicates that at this stage a plastic hinge develops in the frame at the position that gave the selected λ.

12. This hinge is inserted and the load factor is increased by a small amount, say 10 per cent. For one cycle the axial loads are kept constant at their current value and the search for the next hinge is initiated once again from step 6. The contribution of a member with a hinge to the overall stiffness matrix is now given by equations (6.14) or (6.15) depending on which end of the member the hinge is located. Whenever a hinge is inserted in the frame the determinant of the overall stiffness matrix with the new hinge included is calculated. The analysis is terminated when this becomes negative, indicating that the frame has collapsed.

6.6. THE NATURE OF THE ITERATION

Figure 6.4(a) represents diagrammatically the variation of the maximum elastic bending moment, in the frame of Figure 6.4(b). This is assumed to occur at point X. The frame has already developed two other plastic hinges at Z and V as shown in the illustration. The last hinge developed at a load factor λ and this is represented by point O in Figure 6.4(a).

Increasing the load factor by a small amount λ^1, say $\lambda^1 = 0.1\lambda$,

14*

and assuming the axial loads are unaltered, the bending moment M_1 at X can be calculated. The axial loads at the load factor $\lambda + \lambda^1$ can also be calculated. This state is represented by point a in Figure 6.4(a). If the applied loads are now increased to a stage where a plastic hinge develops at X, a linear extrapolation locates point b, where the bending moment at X is equal to M_p and the new load factor is $\lambda + \lambda^{11}$. However, if the axial loads in the members are also calculated by a linear extrapolation between points O and a

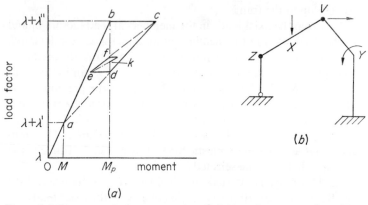

Figure 6.4. Diagrammatical representation of the iteration process for a hinge
(a) Iteration for a hinge at X
(b) Position of hinges in a frame

and used in the analysis of the frame at a load factor $\lambda + \lambda^{11}$, then the bending moment at X will be greater than the plastic moment. This is represented by point c. A linear interpolation between a and c locates point d giving the load factor that brings the bending moment at X back to M_p. Again by reducing the axial loads in the members in the same manner, the bending moments at X will be different from M_p. This is represented by point e. Repeating this process the iteration finally settles down at a point such as k with the bending moment at X equal to the plastic moment.

Care must be taken, throughout the process, that the bending moment at no other point becomes more than the value at X. If it does, at Y say, then the iteration process aims at making the bending moment at Y converge to M_p.

196

6.7. ITERATION TO CONSTANT M_p

Generally for frames with low axial forces in the members, the effect of these forces on the full plastic moment M_p of the sections is small and can be neglected. In such frames, when the bending moment in a member reaches the value M_p, a hinge may be inserted in that member. Consider the bending moment at one end of a particular member to have been M_1 at the previous cycle of iteration and

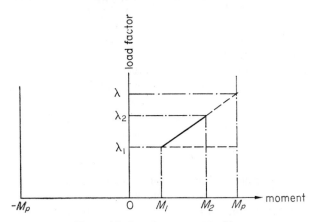

Figure 6.5. Iteration to constant Mp

to be M_2 at the current cycle. Let the corresponding load factors be λ_1 and λ_2 as shown in Figure 6.5, the fully plastic moment of the member being M_p. As the load factor is increased the bending moment tends towards a positive or a negative M_p. From the figure, this sign is determined by the sign of the ratio $(\lambda_2-\lambda_1)/(M_2-M_1)$. When this ratio becomes negative the bending moment approaches a negative M_p value.

The linear prediction for the load factor λ at which the full plastic moment of the section will be reached is, by similar triangles, given by either of the following two equations

$$\lambda = \lambda_2 + \frac{M_p - M_2}{M_2 - M_1}(\lambda_2 - \lambda_1); \qquad (6.29a)$$

or

$$\lambda = \frac{\lambda_2 - \lambda_1}{M_2 - M_1} M_p + \frac{\lambda_1 M_2 - \lambda_2 M_1}{M_2 - M_1} \qquad (6.29b)$$

197

Although the value of λ is the same whichever equation is used, it is noticed that when M_p nearly equals M_2, rounding off errors may change the sign of $M_p - M_2$ and thus equation (6.29a) can lead to a wrong prediction of λ. Equation (6.29b) on the other hand does not suffer from such a defect and can therefore be adopted.

The axial load p_λ in a member corresponding to the load factor λ is also calculated by linear extrapolation from

$$p_\lambda = \frac{\lambda - \lambda_1}{\lambda_2 - \lambda_1} p_2 - \frac{\lambda - \lambda_2}{\lambda_2 - \lambda_1} p_1 \qquad (6.30)$$

where p_1 and p_2 are the axial loads in the member at the load factors λ_1 and λ_2 respectively.

It was pointed out that, from Figure 6.5, if $(\lambda_2 - \lambda_1)$ and $(M_2 - M_1)$ have different signs, the bending moment at the end of the member tends towards $-M_p$. However, the case may arise when the values of M_1, M_2 and M_p are nearly equal. This can happen when several hinges are about to form at once, in which case it is very likely that, due to rounding off errors, a wrong prediction will be made. For this reason it is necessary to compose the product of $M_1 M_2$ with M_p^2 and if they are within a specified tolerance, then the values of M_1, M_2 and M_p are all nearly equal and the sign of M_2 is allocated to M_p. In this manner if both M_1 and M_2 are positive, the iteration is carried out towards M_p. If they are both negative the iteration is towards $-M_p$. If M_1 and M_2 have different values or a different sign then a wrong prediction cannot take place, since the above tolerance test will show that M_1 and M_2 are different.

Once a hinge is inserted at the end of a member the values of M_1 and M_2 remain equal to M_p at two successive cycles of iteration. This criterion may be employed to exclude that end of the member from being selected once more for the iteration purpose. However in large frames, when more than one hinge is being developed at a given load parameter, these are inserted one at a time. After the insertion of one hinge the predicted value of λ for the next hinge remains unaltered, and thus the bending moments remain equal to M_p for two cycles of iteration. This leads to the breakdown of the above criterion. For this reason it is necessary to inspect the ends of the members meeting at a joint and select those without a hinge for iteration, unless, of course, the member selected is the last one at the joint without a hinge. For a joint where n members meet $n-1$ hinges can develop. This method of inspection is necessary when a computer is used for the analysis.

6.8. ITERATION TO REDUCED M_p

For structures with high axial loads in the members, such as tall multi-storey frames, the full plastic moments of these members are reduced appreciably. The variation of the plastic hinge moment of a member with respect to the load parameter of the structure and the axial load in the member is shown by curve ABC in Figure 6.6. The

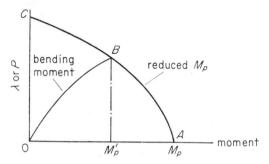

Figure 6.6. The variations of bending moment and the plastic hinge moment at a point with the load factor and plastic axial load in the member

variation of the actual bending moment in the member is also shown by curve OB. It is observed that, with λ increasing, both the bending moment and the axial load increase while the plastic moment of the section decreases. This process continues until at B, the curve OB intersects ABC and a plastic hinge is developed with the reduced plastic hinge moment of the section being M_p^1. Further increase in the load parameter causes a reduction in the bending moment and the plastic hinge moment of the section, as shown by the portion BC of the graph. For British standard beams and Universal Beams and Columns under axial loads, Horne[10] derived the reduced plastic modulus S^1 of a section to be given by:

$$
\text{and} \quad
\left.
\begin{aligned}
S^1 &= S - Fn^2 && \text{for} \quad n < n^1 \\
S^1 &= G(1-n)(n^{11}+n) && \text{for} \quad n > n^1
\end{aligned}
\right\}
\qquad (6.31)
$$

where S is the plastic section modulus of the unstressed section, n is the ratio of the axial load to the squash load of the section i.e. the product of the yield stress and the area of the section, n^1 is the value of n when the web of the section is fully yielded, and G, F and n^{11} are section constants given in standard tables of section properties.

Because of the continuous reduction of the plastic hinge moment of the section the linear extrapolation of Section 6.7, Figure 6.5, overestimates the predicted value of λ. In Figure 6.7 a refined method is represented for the prediction of the load factor λ at which a hinge forms at an end of a member. In this figure points C and B represent the bending moments M_1 and M_2, at one end of the member at two successive cycles of iterations at load factors λ_1 and λ_2 respectively. At these load factors the reduced plastic hinge moments M_{p1}^1 and M_{p2}^1 are given by points D and E. As the load factor

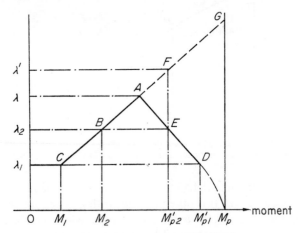

Figure 6.7. Iteration to reduced Mp^1

is increased, the straight lines CB and DE meet at A and a plastic hinge is developed in the member.

The iteration towards a constant M_p locates point G in the figure and an improved prediction can be obtained by iterating towards M_{p2}^1, the reduced plastic moment of the section at load factor λ_2 as shown by point F of Figure 6.7. In doing so, equation (6.30) can be used provided M_{p2}^1 replaces Mp in the formula. However iterating towards a wrong point F instead of A increases the number of cycles of iteration. Furthermore wrong prediction of λ^1 instead of λ gives rise to oscillation. For instance, when a prediction is made of the load factor at which a hinge is to form in a column and λ is increased to this value, the value of M_p^1 is reduced, necessitating a further iteration. Oscillation can become a real problem when more than one hinge is about to form, and particularly when there is the likelihood of hinges forming in beams as well as columns. Oscillation can in

200

some cases be prevented with the use of a larger tolerance on the load factor, leading to a less accurate analysis.

In order to reduce computation and eliminate oscillation, a refined procedure of predicting the ordinate λ of point A in Figure 6.7 is recommended. In this case the value of λ is obtained from the intersection of the lines CBA and DEA. Thus, for a hinge to develop at the end of a member, the value of λ is given by

$$\lambda = \lambda_1 + \frac{(\lambda_2 - \lambda_1)(M_{p1}^1 - M_1)}{M_{p1}^1 - M_{p2}^1 + M_2 - M_1} \tag{6.32}$$

The axial loads in the members are calculated using equation (6.30) and the reduced plastic hinge moments from equations (6.31).

6.9. ALTERATIONS TO THE LOAD VECTOR

It was shown in Section 6.3 that the load vector corresponding to the overall stiffness matrix of a frame has $(3m+n)$ elements where the frame has m joints and n hinges. As the load factor is altered, the first $3m$ elements of the load vector are changed accordingly so that each element is equal to the product of its initial value, at the load factor of unity, and the current load factor. The last n elements of the load vector are the hinge moments. For real hinges the hinge moment is kept at its initial zero value.

For each plastic hinge in a member the axial load in the member at the current load factor is used to obtain the reduced plastic hinge moment of the section. This plastic hinge moment is then used as an element of the load vector. As each plastic hinge is developed in the frame, the order of the load and displacement vectors is increased by one element while the overall stiffness matrix of the frame gains a new row and column. Thus the size of the problem to be solved is continuously increasing as collapse is approached.

6.10. A COMPUTER PROGRAMME FOR THE ELASTIC-PLASTIC ANALYSIS OF GENERAL FRAMES

The programme described here is for plane rigidly-jointed frames with or without real hinges. The iteration process is carried out towards a reduced M_p^1 value given by point A of Figure 6.7.

It is assumed that once a plastic hinge is developed it will continue to rotate in the same direction until the frame collapses. During an elastic-plastic analysis, as the load factor increases, it is possible for a plastic hinge to stop rotating and become inactive. The manner in which the analysis is altered for the treatment of inactive hinges is given later in Section 6.12.

The data for the programme and the manner in which the overall stiffness matrix is constructed are similar to those for the non-linear elastic analysis programme given in Section 4.6. However, the iteration process for an elastic-plastic analysis is different, since the load factor is no longer constant. This process and the pro-

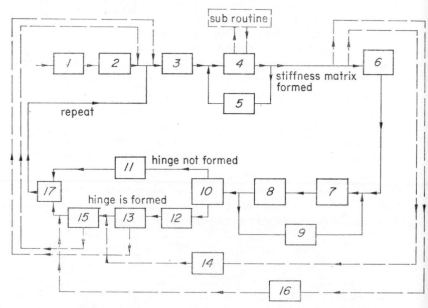

Figure 6.8. Flow diagram of a programme for the elastic-plastic analysis of frames

gramme procedure are best described with the aid of the flow diagram of Figure 6.8. The different stages are numbered in the diagram and these are

1. (a) Read modulus of elasticity, number of members in frame and member data. Read number of joints and real hinges in the frame.

   (b) Set initial value of axial loads to zero.

2. (a) Read load vector L and store on two different locations.

   (b) Read tolerance.

   (c) Read value of full plastic moment of each member together with constants of equations (6.31) for the reduction of plastic hinge moments.

3. Clear a space in the store for the stiffness matrix.

4. Take one member at a time.

   (a) Test if axial load is zero. If so, set the stability functions $\phi_1 - \phi_5$ to unity or else calculate these functions.

   (b) Test if end 1 of the member is fixed. If not, calculate the contribution of this end to the stiffness matrix. Enter subroutine to add these contributions row by row to their locations in the store.

   (c) Test if end 1 has a hinge. If so, enter subroutine to add the contribution to the column of the hinge rotation θ_H.

   (d) Repeat steps (a), (b) and (c) for end 2 of the member.

5. Repeat for next member.

6. (a) Solve the set of simultaneous equations $L = KX$ for joint displacements X. Print these if required.

   (b) Store hinge rotations.

7. (a) For each member calculate the stability functions.

   (b) Calculate new axial load in the member.

   (c) Calculate the bending moment at each end of the member.

   (d) Calculate the reduced plastic hinge moment of each section using the axial loads of step b and equations (6.31).

8. (a) Test if a hinge can be added to end 1 of the member.

   (b) If so, predict the load factor λ to make the bending moment at this end equal to M_p^1. Preserve the lowest λ.

   (c) Repeat steps a and b for the second end of the member.

9. (a) Repeat from step 7(a) for all the other members. Save the lowest value of λ and identity of the corresponding member, i.e. its number and its end number, at which a hinge might take place with this value of λ.

(b) Print the information gained in step (*a*) for reference.

10. Test if λ is within the tolerance of its previous value.

11. (a) If λ is not within the tolerance of its previous value, factor the elements of the load vector, extrapolate the values of the axial loads and calculate the reduced plastic hinge moments of the members due to the newly calculated member forces.

(b) Enter the M_p^1 values of the hinges into the load vector.

12. If λ is within the tolerance of the previous value then a hinge is formed. Therefore print the position of the hinge, the deflections at the joints and the values of the axial loads at the present and previous rounds of iteration. These may be used as a check on the accuracy of the analysis.

13. Test if hinges are all active.

14. If a hinge is not active calculate the determinant of the stiffness matrix and print its value. Stop the programme.

15. If all hinges are active.

(a) Write the appropriate hinge number h_1 or h_2 i.e. the hinge parameter of the member where the hinge has formed, to the member data.

(b) Increase the size of the stiffness matrix by an extra row and column.

(c) Add the value of M_p^1 of the member to the end of the load vector.

(d) Factor λ and the joint loads by a constant small amount.

16. Calculate the determinant of the stiffness matrix and print its value. Stop if this is negative.

17. Enter the load vector into both storage locations. Repeat from step 3.

The programme just described uses the overall stiffness matrix shown in Figure 4.11. As a result of solving the stiffness equations the calculated elements of the displacement vector overwrite the corresponding elements of the load vector. It is for this reason that the load vector is stored at two different stores. When one of these vectors is overwritten, the other one is available to be used in the iteration process.

Throughout the analysis the stability of a frame is checked by calculating the determinant of the overall stiffness matrix. The frame collapses when this determinant becomes negative and the analysis is terminated. At the point of collapse, the frame is at the state of neutral equilibrium, and it is therefore possible for it to reverse the direction of its deflection. This may lead to the reversal of hinge rotations. It is for this reason that once an inactive hinge is recorded, the determinant of the overall stiffness matrix is recalculated. By the sign of this determinant it will be possible to predict whether the inactive hinge is due to neutral equilibrium at collapse or whether it is a genuine one.

From Figure 4.11 it is clear that rows and columns corresponding to the hinges are added below and to the right of the rows and columns of the original stiffness matrix. This method of accommodating the contribution of the hinges has no disadvantages when using a full matrix storage scheme. Later on in this chapter the properties of the stiffness matrix are studied in further detail so that they may be used to economise computer time and storage.

6.11. WORKED EXAMPLES

The first example, in the use of the above computer programme, is the two-storey frame of Figure 1.10 previously discussed in Section 1.4. The dimensions and member properties are given in

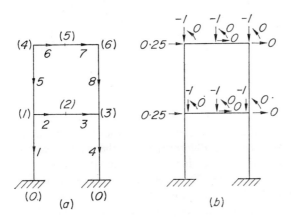

Figure 6.9. Schematic diagram for the frame of Figure 1.10
(a) Schematic diagram   *(b) Loading at $\lambda = 1$*

that section. Figure 6.9(a) shows the numbering of the members and joints, and Figure 6.9(b) shows the loading for $\lambda = 1$. As the frame has six joints and no real hinges the initial load matrix to be specified to the computer has 18 elements as follows

$$L = \{0{\cdot}25 \ -1 \ 0 \ 0 \ -1 \ 0 \ 0 \ -1 \ 0 \ 0{\cdot}25$$

$$-1 \ 0 \ 0 \ -1 \ 0 \ 0 \ -1 \ 0\}$$

In Figure 6.10 the non-dimensional horizontal deflection Δ/L of joint 6 is plotted against λ, alongside a corresponding curve obtained manually. The deformed shape of the frame at the collapse load,

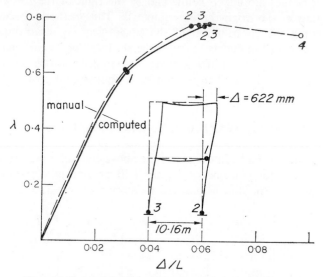

Figure 6.10. Load-deflection diagram for frame of Figure 6.9

with $\lambda = 0{\cdot}7778$, is also shown in the figure together with the final horizontal deflection at the top beam level. After the third hinge was introduced, the determinant of the overall stiffness matrix became negative, indicating an unstable structure. A total of fourteen solutions of the stiffness equations were required in performing the whole analysis. A comparison of the graphs in Figure 6.10 shows good agreement between the manual and the computer analysis, both for the behaviour of the frame up to collapse and the ultimate carrying capacity of the frame.

The results of a computer analysis of the frame shown in Figure 6.11(a) are given as a second example. The instability effects in this tall and slender frame are aggravated by supporting it on horizontal beams to simulate differential settlement of the founda-

206

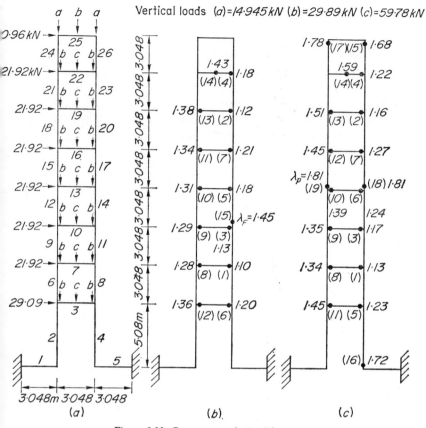

Figure 6.11. Computer analysis of frame

tion when the frame is subject to wind loading. The dimensions and the working loads at unit load factor and the plastic section modulii of the members are shown in Figure 6.11(a). The frame consists of Universal Beams and Columns. These sections are given in Table 6.1.

In Figure 6.11(b) the positions and order of formation of the hinges are given together with the load factors at which they form. In this analysis the effect of axial loads in reducing both the stiffness of the frame and the plastic hinge moments of the members was considered. In a subsequent analysis the axial load effects on the stiffness of the frame were neglected while their effects on the plastic hinge moments were taken into consideration. The result of this piecewise linear elastic-plastic analysis is shown in Figure 6.11(c). It is noticed that the instability effects of the axial loads change the order of hinge formation as soon as the fifth hinge is developed. The hinges also develop at lower load factors and final collapse takes place with fifteen hinges at a load factor of 1.43. Neglecting the stability effects of the axial loads result in collapse by a mechanism with nineteen hinges at a load factor of 1.81 which is 26.6 per cent higher than the first analysis. The difference in the mode of collapse is also apparent in the figures.

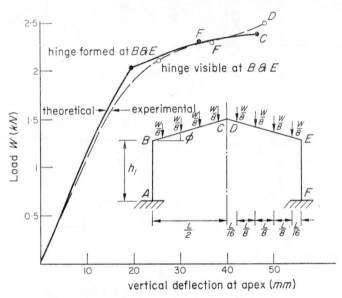

Figure 6.12. Load-deflection curve for a pitched roof frame

As a final example, the load deflection curves for the frame in Figure 6.12, obtained both by the computer and experimentally, are compared. The frame is loaded symmetrically as shown in the figure. It has a span of 1219 mm, column height of 406 mm and an angle of pitch of 15°. The members were manufactured out of 13 mm × 13 mm square black mild steel bars which had a full plastic hinge moment of 153·793 kNmm. EI was 460748 kNmm². The frame collapsed with four hinges at B, E, F and D at a load of $W = 2·469$ kN. The theoretical collapse load was $W = 2·406$ kN which is 2·5 per cent below the experimental results. This difference is mainly due to the strain hardening effect of the

208

Table 6.1

Member	Section
Beams	
1,5	$36 \times 16\frac{1}{2} \times 230$ U.B.
3	$16 \times 7 \times 36$
7	$14 \times 6\frac{3}{4} \times 34$
10	$14 \times 6 \times 34$
13	$14 \times 6\frac{3}{4} \times 30$
16	$12 \times 6\frac{1}{2} \times 27$
19	$10 \times 5\frac{3}{4} \times 21$
22	$8 \times 5\frac{1}{4} \times 20$
25	$8 \times 5\frac{1}{4} \times 17$
Columns	
2,4	$14 \times 16 \times 158$ U.C.
6,8	$14 \times 14\frac{1}{2} \times 87$
9,11	$12 \times 12 \times 79$
12,14	$10 \times 10 \times 49$
15,17	$8 \times 8 \times 35$
18,20	$8 \times 8 \times 31$
21,23	$8 \times 8 \times 31$
24,26	$6 \times 6 \times 15 \cdot 7$

material which was visible as the plastic hinges were spreading along the members. The load deflection graphs show good agreement between the two results. It is noticed that due to strain hardening, the slope of the load-deflection graph, obtained experimentally, increases after the formation of the third hinge at F.

6.12. INACTIVE HINGES

It was stated in Section 6.10 that it is possible for a plastic hinge to develop in a frame, and to continue to rotate, as the load factor increases and then stop rotation and become inactive. For instance in the case of a beam subject to several point loads, it is possible for a hinge to develop under one of the point loads but to stop later while a hinge develops under a different point load.

The matrix partitioning method given by equations (4.14) and (4.15) of Section 4.4 can be used, with advantage, in order to overcome the difficulty introduced by inactive hinges. Consider the case where after the development of several hinges in a frame,

s_{ome} of these hinges have become inactive. When a hinge becomes inactive the end of the member where the hinge was located, becomes elastic with the hinge leaving a permanent kink in the member. This permanent distortion is equal to θ_H which is the total hinge rotation before it ceased to exist. The values of these hinge rotations are known and are given in the displacement vector obtained by solving the equations $L = K.X$ under the current load factor. All these hinge distortions for the inactive hinges can be collected together. Let the remaining degrees of freedom in the frame be termed X_1 and the known inactive hinge distortions be X_2. The load vector corresponding to X_1 is L_1 which includes the actual joint loads together with the M_p^1 values of the remaining active plastic hinges. The loads corresponding to X_2 are L_2, the values of which are not important in the calculations. The set of equations $L = K.X$ can be partitioned so that

$$\begin{bmatrix} L_1 \\ L_2 \end{bmatrix} = \begin{bmatrix} a_{11} & a_{12} \\ a_{21} & a_{22} \end{bmatrix} \begin{bmatrix} X_1 \\ X_2 \end{bmatrix} \tag{6.33}$$

where

$$K = \begin{bmatrix} a_{11} & a_{12} \\ a_{21} & a_{22} \end{bmatrix} \tag{6.34}$$

The first set of equations (6.33) gives

$$L_1 = a_{11}X_1 + a_{12}X_2$$

Hence

$$X_1 = a_{11}^{-1}\{L_1 - a_{12}X_2\} \tag{6.35}$$

Thus the unknown displacements X_1 are calculated from equations (6.35). The elements of the right-hand side of all the matrices of these equations are known. The rest of the elastic-plastic analysis is continued in the usual manner. Further details concerning the alteration of the stiffness matrix to carry out the above analysis with inactive hinges are given by Davies[11].

6.13. COMPUTATIONAL ECONOMY

When analysing the two-storey frame of Figure 6.9, Section 6.11 it was stated that fourteen solutions of the stiffness equations were required to locate three hinges in the frame. When using the programme outlined in Section 6.10 it was found that between two and five iterations were normally needed to locate each hinge when the tolerance was set at 0·0001. For very large frames this may result in a considerable consumption of computer time. Furthermore, if a full matrix scheme is used the available computer store may limit the size of frames that can be analysed. For these reasons it is necessary to make use of all the special features of the stiffness matrix so that both computer time and storage are used economically.

The contributions of a member, connecting joints i and j, to the overall stiffness matrix of a frame, are given in equations (6.14) and (6.15). The contribution of other members is similar to member ij and when more than one member is connected to a particular joint the contributions of these members are accumulative. For most frames each joint is directly connected to only a small member of the remaining joints in the frame. For these reasons the stiffness matrix has the following particular features

(i) That a large number of its elements are zero;
(ii) That the elements occupy a band of irregular boundary;
(iii) That the overall stiffness matrix is symmetrical as shown in equations (6.14) and (6.15).

The non-zero submatrices for a three-storey, two-bay frame, for instance, are shown shaded in Figure 6.13 where the irregular

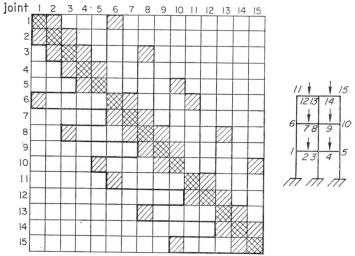

Figure 6.13. The overall stiffness matrix for a 3-storey 2-bay frame

boundary of the non-zero submatrices is shown by a thick line. All the elements outside the boundary line are zeros and need not be stored or operated upon. Furthermore, because the stiffness matrix is symmetrical only the elements on and to one side of the leading diagonal need to be stored and operated upon. Making use of these features of the stiffness matrix and using the notations of equation (2.41a) for the submatrices of a member, then the storage required by the overall stiffness matrix is evaluated as follows: Each 3 by 3 submatrix of Figure 6.13 consists of 9 elements. If i is the lowest numbered joint that is connected to a given joint j, where $j > i$ then $9(j-i)$ locations are required to store the K_{ig} submatrices, where $g = i, i+1, i+2, \ldots j-1$. Since only the elements to one side of the leading diagonal are being stored K_{jj} requires therefore a further six locations, and the total storage N required in the elastic analysis of a frame with M joints is given by

$$N = 6M + \sum_{j=1}^{j=M} 9(j-i) \tag{6.36}$$

A further $3M$ locations are also required to store the address locations of the leading diagonal elements for identification purposes. For the frame of Figure 6.13, for instance, four hundred and seventy-seven locations are required to store the stiffness matrix as compared with two thousand and twenty-five locations needed for a full matrix operation. Equation (6.36) demonstrates how the storage requirement is governed by the manner in which the joints of a structure are numbered for as $j-i$ increases, so also does N. In Figure 6.13 it is noticed that over a third of the locations within the irregular boundary are also empty.

During the construction of the stiffness matrix the rows corresponding to the hinges may be inserted below those of the original matrix of the elastic structure as shown in Figure 4.11. However the figure shows that when the joint numbers i and/or j at the ends of a member are low, a large number of empty locations are wasted in the rows corresponding to the hinges. In order to save computer time and storage space, the hinge rows are therefore moved just below the rows of the corresponding joint.

Figure 6.14 shows part of the overall stiffness matrix constructed in this manner for the first three joints of a structure, together with three hinges near joints 1 and 3. From the figure it is seen that the first row of a joint j in the stiffness matrix would be

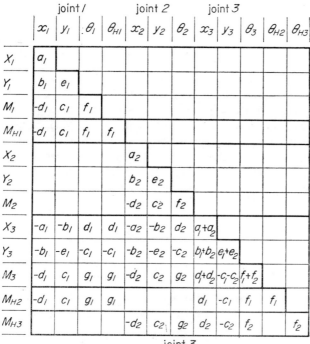

	joint 1				joint 2			joint 3				
	x_1	y_1	θ_1	θ_{H1}	x_2	y_2	θ_2	x_3	y_3	θ_3	θ_{H2}	θ_{H3}
X_1	a_1											
Y_1	b_1	e_1										
M_1	$-d_1$	c_1	f_1									
M_{H1}	$-d_1$	c_1	f_1	f_1								
X_2					a_2							
Y_2					b_2	e_2						
M_2					$-d_2$	c_2	f_2					
X_3	$-a_1$	$-b_1$	d_1	d_1	$-a_2$	$-b_2$	d_2	a_1+a_2				
Y_3	$-b_1$	$-e_1$	$-c_1$	$-c_1$	$-b_2$	$-e_2$	$-c_2$	b_1+b_2	e_1+e_2			
M_3	$-d_1$	c_1	g_1	g_1	$-d_2$	c_2	g_2	d_1+d_2	$-c_1-c_2$	f_1+f_2		
M_{H2}	$-d_1$	c_1	g_1	g_1				d_1	$-c_1$	f_1	f_1	
M_{H3}					$-d_2$	c_2	g_2	d_2	$-c_2$	f_2		f_2

$$a = (EA \cos^2 \alpha + 12EI\, \phi_5 \sin^2 \alpha / L^2) / L$$
$$b = (EA - 12EI\, \phi_5/L^2) \sin \alpha \cos \alpha / L$$
$$c = 6EI\, \phi_2 \cos \alpha / L^2$$
$$d = 6EI\, \phi_2 \sin \alpha / L^2$$
$$e = (EA \sin^2 \alpha + 12EI\, \phi_5 \cos^2 \alpha / L^2) / L$$
$$f = 4EI\, \phi_3 / L$$
$$g = 2EI\, \phi_4 / L$$

Figure 6.14. *Stiffness matrix for part of a frame consisting of two members, three joints and three hinges*

given by

$$r_{j1} = 3j - 2 + \sum_{i=1}^{j-1} f_i \tag{6.37}$$

where f_i is the total number of hinges round joint i. The summation term for $j = 1$ is obviously disregarded. On the other hand the

213

row corresponding to a hinge is given by

$$r_h = 3j + \sum_{i=1}^{j-1} f_i + h \qquad (6.38)$$

where h is the order of the hinge at joint j. Hinges at each joint are counted from one upwards, starting with the hinge in a member connecting j to the joint with the lowest number, and continuing in ascending numerical order of joints as shown in Figure 6.14. Details of computer programmes for non-linear elastic and elastic-plastic analysis of large frames, making use of these features, are given by Majid and Anderson[12]. The programmes make use of a compact storage scheme due to Jennings[13].

EXERCISES

1. Construct the overall stiffness matrix for the pinned base pitched roof frame shown in Figure 6.15. Make the order of this matrix as small as possible. The frame is symmetrical and loaded

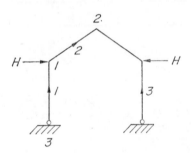

Figure 6.15

by equal horizontal forces H at the eaves. Where does the first plastic hinge develop?

Answer.

$$\begin{bmatrix} b_1 & -d_1 & -d_1 \\ -d_1 & e_1+e_2(1+c_2^2) & f_1 \\ & -2c_2f_2 & \\ -d_1 & f_1 & e_1 \end{bmatrix}, \quad X = \{x_1 \ \theta_1 \ \theta_3\},$$

at the head of columns 1 and 3.

214

2. Calculate the elastic-plastic failure load of the portal frame in Figure 1.4, whose section properties are given in Section 1.3. Take the effect of axial loads into consideration. What is the order of plastic hinge formation?

Answer. $P = 113 \cdot 34$ N, F, C, D.

3. The symmetrical fixed base pitched roof frame in Figure 6.16 is loaded by an increasing vertical load W acting at the apex. Carry out an elastic-plastic analysis of the frame up to failure

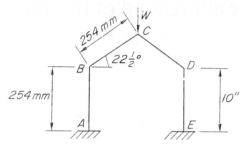

Figure 6.16

indicating the order of hinge formation and the value of W_F at failure. Take $E = 207$ kN/mm², $M_p = 9 \cdot 27$ kNmm. The section is rectangular with $b = 12 \cdot 7$ mm and $d = 3 \cdot 175$ mm

Answer. C, B and D, A and E. $W_F = 195 \cdot 45$ N.

CHAPTER 7

Failure load of frames (Approximate methods)

7.1. INTRODUCTION

In the previous chapter the fundamental principles were given relating to the elastic-plastic analysis of frames. A detailed computer analysis of frames up to failure was described together with factors affecting the accuracy of the analysis or the economy of the comutation.

It is clear that for the simplest frame the calculation of the failure load by an elastic-plastic analysis is lengthy and the use of a computer is necessary. Such an analysis, however, is necessary not only to predict the failure load factor λ_F but also to investigate exactly where, during the process of loading, the plastic hinges develop.

Nevertheless on many occasions, the engineer is interested in a quick and approximate estimate of the carrying capacity of a frame without an elaborate elastic-plastic analysis. For this reason two such methods are given in this chapter. For many small frames these can be used to evaluate the failure load with reasonable accuracy. Later on in the chapter, the results obtained by these methods are compared with those obtained by experiments. In this way the merits of each method are assessed.

7.2. THE RIGID-PLASTIC DROOPING CURVE

In Section 1.4 it was seen that the effect of axial loads in the members of a frame alters its rigid plastic behaviour by modifying the straight line DE of Figure 1.9 to the drooping curve DE^1. This is because as soon as the rigid plastic collapse takes place at point D, Figure 1.9, the frame begins to deflect and the axial loads do internal work that has to be included in the virtual work equation. For the combined mechanism of Figure 1.7(c), for instance, it was stated that the work equation alters from that given by equation (1.8) to that given by equation (1.16). It is noticed from equation (1.16) that, because of the axial loads, the external loads P and $P/4$ have to be reduced in order to balance the work absorbed by the hinge rotations. For this reason the ordinate of any point on the drooping curve is less than the rigid plastic load factor.

Apart from its initial steepness the drooping curve flattens out and generally, at collapse, or soon after, the elastic-plastic curve $OG^1H^1I^1FJ^1$, Figure 1.9, and the drooping curve merge. These facts can be utilised to select a point on the drooping curve which is very near the actual failure load of the frame given by point F. For this purpose the derivation of the equation of the drooping curve is necessary. Consider first a simple cantilever column AB of Figure 7.1 which is subject to a lateral load λP at A together with an axial load $r\lambda P$. According to the rigid plastic theory, there will be no

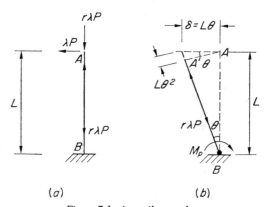

Figure 7.1. A cantilever column

(a) Dimension and loading

(b) Displacements

deformation in the member until collapse takes place at $\lambda = 1$ and a plastic hinge develops at B. After collapse, the hinge begins to rotate and at a given value θ of this rotation, the work absorbed by the hinge is $Mp\theta$. The work done by the horizontal force is $\lambda PL\theta$, where $L\theta$ is equal to the horizontal movement δ of point A. At the same time the work done by the axial load is $r\lambda PL\theta^2$ where $L\theta^2$ is the small distance along which the axial load travels, as shown in Figure 7.1(b). The work equation for the post collapse condition therefore becomes

$$M_p\theta = \lambda PL\theta + r\lambda PL\theta^2$$

Substituting δ for $L\theta$ and simplyfing this equation gives

$$\delta = (M_p - \lambda PL)/r\lambda P \tag{7.1}$$

which is the equation of the drooping curve. It is noticed that at the instant of the hinge formation, $\lambda = 1$ and $M_p = PL$, the value of δ, from equation (7.1), becomes zero. For values of λ below unity, $M_p > \lambda PL$ and δ is positive.

The above procedure for a member can be generalised and used to derive the equation of the drooping curve for frames. For the combined mechanism of the portal frame of Figure 7.2, for instance, the compressive forces in the members are R_1, R_2 and R_3 as shown

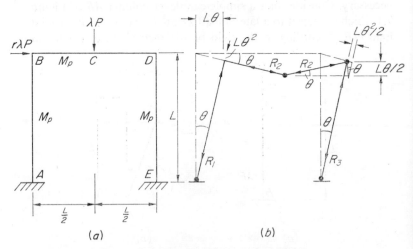

Figure 7.2. Combined mechanism of portal frame
(a) Loads and dimensions
(b) Combined mechanism

in the figure. If the hinge rotation at A is θ, then the work done by R_1 and R_3 are $-R_1L\theta^2$ and $-R_3L\theta^2$ respectively. The work done by the axial load in each part of the beam is $-R_2L\theta^2/2$. The distances through which these forces move are shown in Figure 7.2(b). The hinge rotations at C, D and E are 2θ, 2θ and θ respectively and the total work absorbed by the four hinges is therefore $6M_p\theta$. Finally the work done by the external loads $r\lambda P$ and λP are $r\lambda PL\theta$ and $\lambda PL\theta/2$ respectively. Equating the external and internal work gives the work equation, thus

$$\lambda PL\theta/2 + r\lambda PL\theta = 6M_p\theta - R_1L\theta^2 - R_3L\theta^2 - 2R_2L\theta^2/2$$

Substituting λP for $R_1 + R_3$, this equation gives the horizontal movement $\delta = L\theta$ of the column heads as:

$$\delta = [6M_p - \lambda PL(r + 0 \cdot 5)]/(\lambda P + R_2) \qquad (7.2)$$

which is the equation of the drooping curve for the combined mechanism.

For the case of a sway mechanism with hinges at A, B, D and E the beam BD moves parallel to itself and the axial force in it does no work. The vertical force λP does no work either. The reader can verify that the equation of the drooping curve for a sway mechanism is therefore given by

$$\delta = (4M_p - r\lambda PL)/\lambda P \qquad (7.3)$$

where δ is the sway $L\theta$ of the columns.

The axial forces in the members can be calculated by an elastic or simple plastic analysis of the frame. The difference in the results of these two analyses does not affect the drooping curve appreciably Using a simple plastic approach the axial loads, for the case of a combined mechanism, can be calculated from the conditions of equilibrium. An easy way of deriving these conditions is to cut the frame at C and introduce the shear force V, the thrust R_2 and the bending moment M_p as shown in Figure 7.3. The vertical load λP at C is placed arbitrarily on the right-hand half of the frame.

Taking moments about point D gives

$$VL/2 + M_p = \lambda PL/2$$

and from vertical equilibrium $V = R_1$, thus

$$R_1 = \lambda P - 2M_p/L \qquad (7.4)$$

The axial force R_3 is calculated by considering the equilibrium of the whole frame which gives $\lambda P = R_1 + R_3$

219

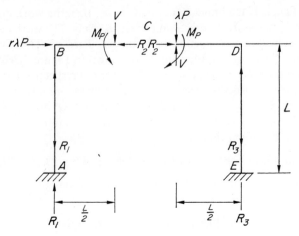

Figure 7.3. Conditions of equilibrium

Hence using equation (7.4)

$$R_3 = 2M_p/L \qquad (7.5)$$

Finally taking moments about point A for the left hand half of the frame

$$M_p + R_2 L = R_1 L/2 + r\lambda PL$$

and using equation (7.4) we obtain

$$R_2 = \lambda P(r+0\cdot 5) - 2M_p/L \qquad (7.6)$$

This value of R_2 can be used in equation (7.2) for the drooping curve of the combined mechanism.

7.3. CALCULATION OF FAILURE LOAD APPROXIMATE METHOD 1

Various load displacement curves for a general frame are given in Figure 7.4. The real load deflection curve of the frame is represented by OFH which reaches a peak value at F where the failure load factor λ_F is given by the ordinate FT. In this figure, $\delta_F(=OT)$ represents a given real deflection in the frame at the state of failure. The first approximate method, proposed by the author[14], calculates the ordinate λ_M of point M instead of the failure load factor. It is

220

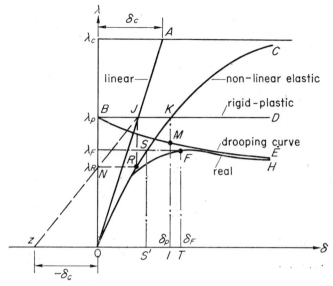

Figure 7.4. Load-displacement curves

noticed that point M is on the drooping curve BME and its ordinate is equal to MI. A third point K, which is vertically above M, is on the non-linear elastic curve $ORKC$ with an ordinate KI which is numerically equal to λ_P given by the simple plastic theory. This point is easily established by a single non-linear elastic analysis, as described in Section 4.2.

The procedure for predicting λ_M therefore consists of the following steps. First the simple plastic load factor λ_p is obtained. At this load factor the elastic deflection δ_p is evaluated taking the full effect of the axial loads into consideration. This deflection and λ_p fix point K on the rigid plastic line BD. Next the work equation is used to derive the equation of the drooping curve. Substituting the calculated value of δ_p in this equation results in the direct evaluation of λ_M.

7.4. ASSUMPTIONS

To accept that λ_M is equal to the failure load factor λ_F depends on the assumption that the elastic deflection δ_p, Figure 7.4, is nearly equal to the real deflection δ_F of the frame at failure. The simple plastic approach ignores the effect of axial loads altogether and assumes

221

that the linear elastic deflection given by BJ in Figure 7.4. is small and can be neglected. For frames where BJ is small and the axial load effects are insignificant then BJ, $BK(= \delta_p)$ and δ_F can all be neglected, resulting in $\lambda_p = \lambda_F = \lambda_M$. These frames collapse with the development of a mechanism. Even if these small deflections are not neglected the values of λ_F and λ_M will still remain early equal to λ_p.

On the other hand frames with excessive axial load effects fail with only a few hinges. Once hinges develop in these frames the deflections increase at a faster rate. At the collapse load factor, therefore, δ_F becomes larger than the deflection given by the non-linear elastic analysis shown by point S in Figure 7.4. However at λ_p, which is larger than λ_F, the member axial loads are higher than those at failure and hence their effect on the elastic deflection is also greater. This suggests that it is reasonable to assume that the contribution of the few hinge rotations on the deflections is of the same order of magnitude as the extra non-linear elastic deflection due to higher axial loads. That is to say, in Figure 7.4., S^1I is nearly equal to S^1T. That $\delta_p \approx \delta_F$ directly follows. Moreover, in practice, the portions ME and FH of the curves BME and OFH are both nearly horizontal. Hence a difference between δ_p and δ_F does not lead to a large difference between λ_M and λ_F.

7.5. ANALYSIS OF COLUMNS

Let us first consider a few individual members and mathematically investigate how close λ_M is to λ_F. The general slope deflection equations used for this purpose are those given by equations (3.31) in terms of n, o and m functions. They are

$$M_{AB} = nk\theta_A - ok\theta_B - SLm/2$$

and

$$M_{BA} = -ok\theta_A + nk\theta_B - SLm/2$$

where S is the shear force in the member. For the case of the cantilever column of Figure 7.1, $S = \lambda P$, θ_B is zero and the bending moment at the free end A is also zero. Thus the first of the slope deflection equations gives

$$k\theta_A = \lambda PLm/2n$$

and therefore the second equation gives

$$M_{BA} = -\lambda PLm(o+n)/2n$$

222

But from the list of the stability functions in Section 3.4, it is noticed that $n+o = 2/m$; it follows that

$$M_{BA} = -\lambda PL/n \qquad (7.7)$$

Now in Figure 7.1, at a given state of deformation, the hogging moment at B due to the external loads is:

$$M_{BA} = -\lambda PL - r\lambda P\delta$$

and equating this value to that given by equation (7.7) results in

$$\delta = L(1-n)/nr \qquad (7.8)$$

which is the equation of the non-linear elastic sway of the column. When the applied load λP is numerically equal to the rigid plastic collapes load then $\lambda_p P = M_p/L$ and the value of the elastic deflection δ becomes equal to δ_p. This is given by equation (7.8), with n_p substituted for n when $\lambda = \lambda p$. Thus

$$\delta p = L(1-np)/npr \qquad (7.9)$$

The deflection δp is shown in Figure 7.4. Point M in this figure is on the drooping curve where the load factor is λ_M. The value of the deflection δp corresponding to this point can also be calculated by using equation (7.1) with λ_M substituted for λ and δp for δ. Thus with these new values of λ_M and δp equations (7.1) and (7.9) give

$$\lambda_M = n_p Mp/PL \qquad (7.10)$$

From this equation the value of λ_M can be calculated directly. Now since $Mp/PL = \lambda p$ it follows that

$$\lambda_M = n_p \lambda p \qquad (7.11)$$

At the failure load factor λ_F, the bending moment M_{BA} reaches the value of the full plastic hinge moment $-Mp$ and from equation (7.7) it follows that

$$\lambda_F = n_F Mp/PL$$

where n_F is the value of the stability function n at $\lambda = \lambda_F$. Now again using λp for Mp/PL we obtain

$$\lambda_F = n_F \lambda p \qquad (7.12)$$

Although $\lambda_F < \lambda_p$, n_F is very nearly equal to n_p. This is because under practical circumstances, the stability function n for a member is not very sensitive to variations in the values of the axial load in that

223

member. In fact, $n_F - n_p \ll \lambda_p - \lambda_F$. Hence the value of $\lambda_M = n_p\lambda_p$ given by equation (7.11) is very close to the failure load $\lambda_F = n_F\lambda_p$ given by equation (7.12). For instance, taking λ_p as unity, for a slender 152 mm$\times$152 mm$\times$69·8 N universal column 6·096 m high, subject to an axial load which is four times the lateral load, i.e. $r = 4$, equation (7.12) gives λ_F as 0·878, while equation (7.11) gives λ_M as 0·864. This represents a difference of only 1·6 per cent. For more realistic columns the difference is even smaller.

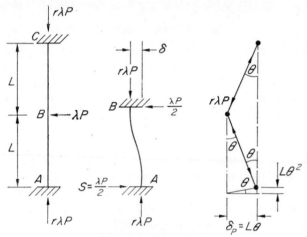

Figure 7.5. Fixed column
(a) Dimensions and loading
(b) Elastic deformation
(c) Mechanism

In the case of the fixed column of Figure 7.5, the elastic lateral displacement at B can be calculated by making use of symmetry and considering only half the member. This is in pure sway as shown in Figure 7.5(b) with only half the lateral load acting at B. For the case of a member in pure sway equation (3.22) can be used to express the shear force $S = 0.5\,\lambda P$ in terms of the sway $v = \delta$ as

$$0.5\lambda P = 2s(1+c)k\delta/mL^2,$$

where m is the stability function $1/\phi_1$. This equation gives

$$\delta = m\lambda PL^2/4s(1+c)k \qquad (7.13)$$

From the list in Section 3.4 it is found that

$$s(1+c) = m\pi^2\varrho/2(m-1).$$

It follows, from equation (7.13) that

$$\delta = \lambda PL^2(m-1)/2\pi^2 k\varrho$$

Now

$$\varrho = r\lambda P/P_E = r\lambda PL/\pi^2 k$$

which gives

$$\pi^2 \varrho k = r\lambda PL$$

Hence

$$\delta = L(m-1)/2r$$

and when the elastic load factor λ reaches the simple plastic load factor $\lambda_p = 4M_p/LP$, δ becomes equal to δ_p, Figure 7.4, and m equal m_p, i.e.

$$\delta_p = L(m_p-1)/2r \qquad\qquad (7.14)$$

The work equation for the plastic collapse mechanism, taking the axial load affects into consideration, Figure 7.5(c), can be derived in the usual manner. This is

$$4M_p\theta = \lambda PL\theta + 2r\lambda PL\theta^2 \qquad\qquad (7.15)$$

In this equation setting λ to λ_M and $L\theta$ to δ_p and using equation (7.14) and the fact that $4M_p/LP = \lambda_p$, we obtain

$$\lambda_M = \lambda_p/m_p \qquad\qquad (7.16)$$

For the case of a member in pure sway the bending moments M_{AB} and M_{BA} are given in terms of the shear force $S = \lambda P/2$ by the first of equations (3.23) as

$$M_{AB} = M_{BA} = -0{\cdot}25\, mL\lambda P$$

At failure the load factor λ reaches λ_F, m becomes m_F and the bending moments M_{AB} and M_{BA} equal $-M_p$. Thus

$$-Mp = -0{\cdot}25m_F L\lambda_F P$$

which upon substitution of $4M_p/PL$ for λ_p gives

$$\lambda_F = \lambda_p/m_F \qquad\qquad (7.17)$$

Once again comparing equations (7.16) and (7.17) since m_p is very nearly equal to m_F, it follows that λ_M is very nearly equal to λ_F. Again this is because the stability function m, like n, is not very sensitive to small changes in the value of ϱ. In fact for realistic columns m is far less sensitive than n.

As an exercise, the reader can calculate the value of λ_M for the propped cantilever column shown in Figure 6.2. Point M for this column is marked on the dropping part C^1D of the non-linear elastic-plastic curve of Figure 6.3 where it is noticed that λ_M is equal to 0·83, a difference of 2 per cent from λ_F.

7.6. ANALYSIS OF FRAMES

The above simple cases demonstrate how close the value of λ_M is to that of the failure load factor. For complex frames it is not feasible to obtain an expression for λ_F to compare with λ_M. This is because the sequence of hinge formation and the number and position of these hinges at collapse vary from frame to frame and a single expression cannot be obtained for λ_F to cover all cases. For this reason attention will instead be given to the general variation of λ_F and λ_M.

In the case of the portal frame of Figure 7.2, the value of λ_M, for the case of a combined mechanism, can be calculated from equation (7.2) of the drooping curve with δ set to δ_p and R_2 calculated from equation (7.6). For a combined mechanism for this frame the value of M_p is given by

$$M_p = \lambda_p PL(r+0·5)/6$$

Using this value in equation (7.6) the axial load R_2 at any load factor λ is given by

$$R_2 = \lambda P(r+0·5) - \lambda_p P(r+0·5)/3.$$

For the particular point with co-ordinates (λ_M, δ_p) on the drooping curve, equation (7.2) now results in

$$\lambda_M = 9\lambda_p M_p/(9M_p + \lambda_p P(2+r)\delta_p) \tag{7.18}$$

On the other hand in the case of a sway mechanism λ_M is calculated from equation (7.3) which gives

$$\lambda_M = 4M_p/(rLP + P\delta_p) \tag{7.19}$$

The value of δ_p in equations (7.18) and (7.19) is calculated by a non-linear elastic analysis at the elastic load factor λ_M. For the purpose of the present, approximate calculation of the failure load λ_M, it is sufficiently accurate to assume that the axial load in the beams can be neglected, while the axial load in each column can be taken as $0.5\,\lambda P$. The non-linear elastic analysis is thus considerably simplified. The reader may use slope deflection equation to find that

$$\delta_p = rL(m-1)(s+12)/(s+12-0.5ms(1+c)) \qquad (7.20)$$

Where m, s and c are the stability functions for a column with an axial load of $0.5\,\lambda_p P$.

For various values of r, the values of λ_M were calculated for a family of frames. These are plotted in Figure 7.6 together with the

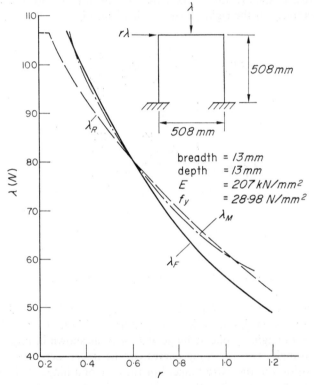

Figure 7.6. Failure load of portals

results of rigorous elastic-plastic analyses following the order of hinge formation. The very flexible frame of Figure 7.6 was selected in order that the curves might be distinguishable. It is evident from the graph that, for values of r up to 0.7, which is very high for any practical loading of a frame, the values of λ_M are very close to λ_F

obtained by following the sequence of hinges. For higher values of r, λ_M becomes larger than λ_F but still in reasonable agreement with it.

As a final example the two-storey portal of Figure 1.10 is considered because it has a number of unusual and interesting features. Taking the general equation for the axial forces into consideration the dropping curve is obtained by writing the work equation for the frame, which upon substitution of δ for $L\theta$, gives

$$\delta = (40M_p - 7\lambda W_p L)/4(\lambda H_1 + \lambda H_2 + 9\lambda W_p) \qquad (7.21)$$

where λH_1 and λH_2 are the axial forces in the top and bottom beams respectively. If the rigid plastic mechanism, Figure 1.10(b), is used

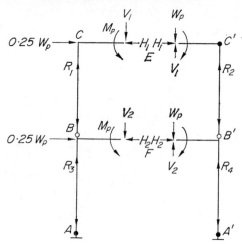

Figure 7.7. Axial forces in the frame of Figure 1.10

to evaluate the forces H_1 and H_2 at $\lambda = \lambda_p = 1$, it is found that the top storey alone is not statically determinate without the insertion of an extra imaginary plastic hinge at B or B^1 as shown in Figure 7.7. Accordingly, two drooping curves can be traced depending upon the position of the extra hinge. For the case of a hinge at B, on the windward column, the values of H_1 and H_2 are given by

$$\left.\begin{array}{l} H_1 = W_p/20 \\ H_2 = 17W_p/40 \end{array}\right\} \qquad (7.22)$$

Using these values in equation (7.21) and equating W_p to $40M_p/7L$, the rigid plastic collapse load at $\lambda_p = 1$, we obtain

$$\lambda = 70L/(70L + 379\delta) \qquad (7.23)$$

228

Alternatively, inserting the imaginary hinge at B^1 gives

$$\lambda = 70L/(70L + 367\delta) \qquad (7.24)$$

For any value of δ, there is very little difference between the two values of λ obtained by either equation. This may be seen in Figure 7.8., where the two equations appear to coincide to give curve B. However, a comparison of equations (7.23) and (7.24)

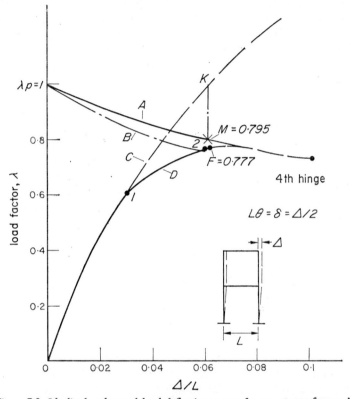

Figure 7.8. Idealised and actual load deflection curves for two-storey frames in Figure 7.7. Curve A, drooping curve obtained by computer; Curve B, drooping curve obtained from equations (7.23) and (7.24); Curve C. nonlinear elastic; Curve D, elastic-plastic

shows that it is safer to calculate λ from equation (7.23). Also in Figure 7.8. the results of a rigorous elastic-plastic analysis are shown by curve D, where it is noticed that collapse takes place with three hinges, as in Figure 1.10(c), at a load factor $\lambda_F = 0.778$. It is also

229

noticed that the drooping curve B actually intersects curve D before collapse takes place. This is an error that comes about by inserting an extra hinge in the frame so that the axial forces can be calculated from the simple plastic collapse mechanism, Figure 7.7. Nevertheless the value of λ_M, M being on curve B vertcally below K, is 0·76, which is 2·2 per cent below λ_F.

In order to establish point K in the figure, a non-linear elastic analysis was carried out at the rigid-plastic load factor $\lambda_p = 1$. The results of this analysis also gave the elastic axial loads in the members. These were used to derive another drooping curve which is shown as curve A in the figure. It appears from Figure 7.8, that curve A merges with the elastic-plastic curve soon after collapse and seems to be more realistic than curve B which passes below the failure load factor. The value of λ_M obtained from curve A is 0·795 which is about 2·2 per cent more than λ_F.

It is also observed that the value of δ_p given by point M or K is very nearly equal to the actual deflection δ_F of the frame at collapse. This indicates that, for this frame, the assumption made that δ_p and δ_F are of the same order, is justifiable. It is noticed how near horizontal is the drooping curve after collapse and for this frame it was found that a difference of ± 20 per cent between δ_F and δ_p would make the values of λ_F and λ_M differ by less than 5 per cent.

7.7. RANKINE'S FORMULA.
APPROXIMATE METHOD 2

This method makes use of the geometrical relationship between the linear and non-linear elastic load deflection curves as given in Section 5.8 and Figure 5.10. In this figure λ^1 is a given load factor at which the linear and non-linear elastic deflections δ^1 and δ_B are calculated. These are represented by points K and B respectively. From these deflections the linear elastic deflection δ_c at the first elastic critical load factor λ_{c1} is calculated using equation (5.52). The elastic critical load factor itself may be calculated using equation (5.53).

Figure 5.10 also has additional uses. A particular point R on the non-linear elastic curve can be established if λ^1 is singled out to be numerically the same as the rigid plastic load factor λ_p. In this case from the similar triangles KHZ and R^1OZ it follows that

$$\lambda_p/\lambda_R = (\delta^1 + \delta_c)/\delta_c = 1 + \delta^1/\delta_c \qquad (7.25)$$

where λ_p and λ_R are shown in Figure 5.10 to be equal to λ^1 and λ respectively. Substituting for δ^1/δ_c from the first part of equation (5.53) into equation (7.25) gives

$$\lambda_p/\lambda_R = 1 + \lambda_p/\lambda_{c1}$$

Hence

$$\frac{1}{\lambda_R} = \frac{1}{\lambda_{c1}} + \frac{1}{\lambda_p} \qquad (7.26)$$

The factor λ_R is called the Rankine–Merchant load factor for a structure. The form of equation (7.26) is similar to the well-known empirical Rankine formula for individual members. Merchant[15] suggested its use to calculate λ_R as an estimate of the failure load factor λ_F. In the form of equation (7.26), the formula is not very useful since it requires the determination of the elastic critical load factor λ_{c1}. The computation involved in this can be considerable, so much so that the advantages claimed for Rankine's formula are overshadowed by the extensive work involved in the determination of the critical load. However, combining equations (5.53) and (7.26) yields

$$\lambda_R = \lambda_p \delta_B/(2\delta_B - \delta^1) \qquad (7.27)$$

In this form, derived by the author[7], the load factor λ_R can be evaluated using the rigid plastic collapse load factor λ_p and the linear and non-linear elastic deflections at this load factor.

The Rankine formula for frames, like its predecessor for individual members, is empirical. λ_R is the ordinate of point R in Figures 5.10 and 7.4. It is noticed that this point is in fact on the non-linear elastic curve. Nevertheless λ_R is considered as a substitute for the elastic-plastic failure load. For frames with excessive side loads, it has been proved by Ariaratnam[16] that the Rankine formula becomes unsafe. However a large number of tests indicate that the formula gives a useful conservative estimate of the failure load. In Figure 7.6 the results of the Rankine formula for a portal are given along with those obtained by rigorous elastic-plastic analyses. It is observed that for values of (r) up to 0·6, λ_R gives safe results. For higher values λ_R becomes unsafe. Nevertheless for all values of (r) shown in the figure, λ_R gives reasonable results. It is interesting to note that for high values of the side load the graphs for λ_R and λ_M nearly coincide.

In Section 5.8 the linear and non-linear elastic sway deflections of the two-storey frame of Figure 1.10 were calculated at roof level. Equation (5.53) was then used with $\lambda^1 = \lambda_p$ as unity. This gave λ_{c1}

as 3.42. Using this value together with $\lambda_p = 1$ in equation (7.26) gives λ_R as 0·775 as compared to λ_F of 0·778, i.e. an error of only 0·38 per cent. The same calculation carried out at the first floor level gave λ_{c1} as 3·32 and λ_R as 0·767.

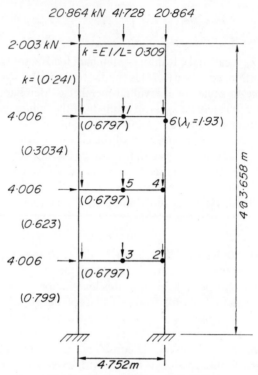

Figure 7.9. Woods four-storey frame

Finally, for the four-storey single bay symmetrical frame of Figure 7.9 the elastic critical load factor λ_{c1} was calculated using the steps of Section 5.3. This gave λ_{c1} as 12·9. A rigid plastic analysis gives λ_p as 2·15 and hence from equation (7.26) λ_R is 1·84. The order of hinge formation, obtained by a rigorous elastic-plastic analysis of this frame, is shown in the figure. The frame fails at a load factor of 1·93 with six hinges. Using equation (5.53) at different storey levels resulted in λ_{c1} varying between 12·8 and 13·4. Nevertheless the corresponding values of λ_R are 1·84 and 1·85 respectively. Thus, all the calculated values of λ_R are within 5 per cent of the actual failure load factor.

7.8. EXPERIMENTAL EVIDENCE

In Table 7.1 the theoretical failure loads of a number of frames are compared with results obtained experimentally. The frames are pitched roof portals with a span L of 1219 mm and loaded as shown in Figure 6.10. Variation in the stability condition of the frames was achieved by altering the column height h, and the pitch angle ϕ. The first three frames were made out of 16 mm×16 mm black mild steel bars machined by 1·6 mm on all faces so that any hard skin that might have developed during the cooling process of the steel was removed. The last four frames were made out of the same batch of 13 mm square bars. The pitch of the first three frames was 22·5° and the others had a pitch of 15°.

The results of a rigorous elastic-plastic analysis are shown in column 4 of the table. These results are generally below those given in column 5, obtained experimentally, with a maximum difference of 6·2 per cent. As was explained in Section 6.11, this is due to the strain hardening effect of the material. The results corresponding to λ_M are shown in column 6 while those corresponding to λ_R are given in column 7. The rigid plastic collapse mechanism used for the derivation of the drooping curve assumed that the four plastic hinges would form at points F, E, B and C, shown in Figure 6.10. A comparison of the last three columns of the table indicates that the results obtained by both approximate methods are in good agreement with the experimental results, the maximum difference being only 5 per cent. These results show that it is reasonable to use either of the approximate method to obtain a quick estimate of the failure load.

EXERCISES

1. The continuous beam $ABCD$ in Figure 7.10 is fixed at A and D and simply supported at C. It is subject to an axial load λP and a point load λP is acting at B. Show that the failure load factor λ_M

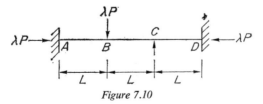

Figure 7.10

is given by.

$$\lambda_M = \frac{\phi_1 \lambda_p}{\phi_1 + (\phi_2 - \phi_5)\left[\dfrac{\phi_4^2 - 16\phi_3^2}{\phi_5\phi_4^2 - 16\phi_5\phi_3^2 + 3\phi_3\phi_2^2}\right]}$$

where $\lambda_p = 4M_p/PL$ is the rigid plastic collapse load factor of the beam. The stability functions are for $\varrho = \lambda_p PL^2/\pi^2 EI$.

2. The continuous beam $ABCD$ in Figure 7.11 is fixed at A and D and simply supported at B and C. It is subject to an axial load λW

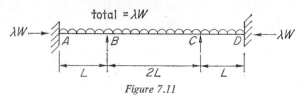

Figure 7.11

and a total uniformly distributed load W. Calculate the Rankine failure load factor of the beam.

Answer.

$$\lambda_R = \frac{9 \cdot 712 \pi^2 EI M_p}{W(0 \cdot 607 \pi^2 EI + 16 M_p L^2)}$$

3. Verify the results given in the first row of Table 7.1 for W_M and W_R for the pitched roof frame shown in Figure 6.12. The frame properties are given in Sections 6.11 and 7.8

Table 7.1 COMPARISON OF THEORETICAL AND EXPERIMENTAL COLLAPSE
LOADS OF FIXED BASE PITCHED ROOF FRAMES

Height to eaves h_1 (m)	Fully plastic Moment (kNmm)	EI kNmm²	Collapse load WkN			
			Elastic Plastic	Experimental	Approximate Method 1	Approximate Method 2
812	147	415	2·17	2·24	2·17	2·24
609	137	363	2·15	2·29	2·18	2·28
406	148	401	2·49	2·64	2·70	2·68
711	153	469	2·20	2·16	2·20	2·26
356	154	506	2·26	2·29	2·30	2·38
508	129	412	1·95	1·93	1·99	1·94
406	154	461	2·41	2·47	2·51	2·50

CHAPTER 8

Elastic-Plastic design of sway frames

8.1. INTRODUCTION

The design procedures practised in structural engineering offices all require the specification of a set of design criteria. The procedure is then to select initial sizes for the members of a given structure, analyse it by means of one of the theories and modify the member sizes accordingly. This process may be repeated until all the design criteria are satisfied. The design criteria themselves are often selected to suit the theory employed in the analysis and design of the structure.

In the case of design by an elastic theory, for instance, a criterion is that the elastic working stresses should not exceed certain permissible values laid down in appropriate specifications such as British Standard 449. For example, provided buckling does not take place, the maximum elastic working stress for steel beams is 162 N/mm² the yield stress of the material being guaranteed to at least 247 N/mm². Thus the factor of safety against yielding anywhere in these beams is 1·524. For universal beams having a shape factor of 1·15, the factor against the formation of a plastic hinge at the most highly stressed point along these beams is thus $1·524 \times 1·15 = 1·75$. For example, a uniformly loaded simply supported beam collapsing with a beam-type mechanism with one hinge, designed by the elastic theory to a permissible working stress of 162 N/mm², has a load factor of 1·75 against collapse.

On the other hand, it can be shown that a fixed-ended similar beam has a load factor of 2·34 against collapse. This indicates that, using the elastic theory, the two beams have the same factors of safety against yield but different load factor against collapse. In general it is true that statically determinate structures, that are designed elastically, have a lower load factor against collapse than statically indeterminate structures. This fact is sometimes considered as a weakness of the elastic design methods.

The simple plastic theory assumes the material of frames to be rigid-plastic, Figure 1.2(b), and, therefore considers that a frame remains unstrained until suddenly, at the simple plastic collapse load factor λ_p a sufficient number of hinges form to convert the frame into a mechanism. Accordingly, using this theory, the design procedure is to proportion the members so that the frame has a definite load factor against collapse. According to BS. 449, since it is permissible for a simply supported beam to have a load factor of 1·75 against collapse, it appears reasonable to adopt this factor for all frames. Furthermore, since elastic working stresses are permitted by BS. 449 to be increased by a quarter, when wind as well as vertical loads are considered, the load factor for frames under combined loading can therefore be reduced to 1·4.

According to the elastic-plastic theory, the hinges in a structure do not all develop at once. For this reason, although load factors against collapse may be adopted which are the same for all frames, the load factor at which individual hinges develop throughout a frame is different from one hinge to another. There is no rational reason to impose the same load factor against the formation of every hinge. In particular, if the permissible load factor for the formation of a hinge is set at that permissible for the collapse of the whole frame, then the entire frame remains elastic up to this load factor; naturally this is not necessary. Indeed it is possible for hinges to develop at any load factor, particularly above the working loads, without collapse taking place below the permissible load factor for collapse.

For these reasons, the use of an elastic-plastic theory as with any other theory requires a new set of design criteria together with a new design procedure. This may appear unfortunate to the designer. Nevertheless, in the case of sway frames, an elastic-plastic approach is necessary not only to predict the failure load factor λ_F, but also to investigate exactly where, during the process of loading, the plastic hinges form. In a report on future research issued by

236

the Civil Engineering Research Council[17] it is stated, under the heading of Design that 'Methods of design must be kept under constant review and attention paid to the broad philosophy and concepts of design as well as to the details. The rapid developments of modern times and the increasing number, size, scope and scale of works will make design much more complex than at present, and it will be necessary in future to design more precisely, taking into consideration secondary and side effects.' In the specific field of sway frames, the problem is complex and the required accuracy demands both a fresh approach to the philosophy of design and greater consideration of secondary and side effects.

In this chapter a summary is given of research carried out on the elastic-plastic behaviour of frames. The design procedure for a computerised method is then outlined and suggestions are made for further improvements in the method.

8.2. VARIATION OF MAXIMUM MOMENTS

Actual variations of the maximum bending moments in members of various frames with the load factor are shown in Figure 8.1. Before the formation of a plastic hinge in a frame, the maximum moments increase with the load factor up to point A, where the first plastic hinge is developed in the frame. The departure from linearity during this stage is due to the effect of axial loads. For higher values of λ the maximum moments generally increase at a faster rate. The curves are plotted for λ up to the permissible load factor at collapse, λ_1. Under combined vertical and wind loading this is usually taken as 1·4.

Curve 1 is for a strong member that remains elastic up to the load factor λ_1. Point B on this curve gives the actual elastic moment in the member at some stage of loading when a hinge develops elsewhere in the frame just above λ_1. Point C, on the other hand, gives the value of the maximum moment in the member at the formation of the previous hinge in the frame at a load factor just below λ_1. Curve 2 is for a member that remains elastic until the frame collapses at a load factor below λ_1. In this case point B gives the final maximum moments in the member at collapse. Curve 3 shows an unusual variation in that the maximum moment in a member begins to decrease after a point shown as D in the figure.

Curve 4 is that of a member where at some stage prior to λ_1, shown by point B, the maximum moment reaches the reduced plastic hinge moment M_p^1 of the member section. At this point a plastic hinge develops in that member. From this stage onwards, any increase in the load factor increases the axial load in the mem-

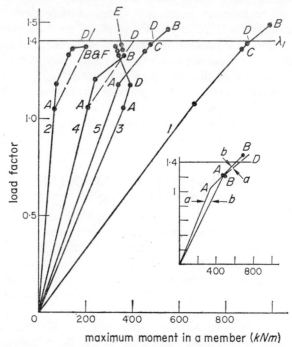

Figure 8.1. Variation of maximum moments with the load factor

ber, and hence the effective plastic hinge moment of the section reduces still further, as shown by portion BE of the curve. Finally, curve 5 is for a member which is strong enough to remain elastic up to the load factor λ_1 but develops a hinge just after that. This is shown by point B. On the same curve point C gives the maximum moment in the member at the formation of a previous hinge in the frame at a load factor just below λ_1.

8.3. DESIGN OF INDIVIDUAL MEMBERS

The variation of maximum bending moments as summarised above and shown in Figure 8.1 can be used for the design of individual members. Suppose that the design criterion is to prevent the formation of a plastic hinge in a member below a load factor λ_1, Beginning with a weak member such as that represented by Curve 4, the linear extrapolation ABD suggests that the section selected for the member should satisfy

$$M_B+(\lambda_1-\lambda_B)(M_B-M_A)/(\lambda_B-\lambda_A) < s^1 fy \qquad (8.1)$$

where fy is the yield stress of the material and s^1 is the reduced plastic section modulus of the section given, for universal beams and columns, by equations (6.31) of Section 6.8. The axial load in the member is calculated from

$$P = P_B+(\lambda_1-\lambda_B)(P_B-P_A)/(\lambda_B-\lambda_A) \qquad (8.2)$$

where P is the axial load at load factor λ_1 and P_A and P_B are the axial loads in the same member obtained during an elastic-plastic analysis at load factors λ_A and λ_B respectively. Subscripts A and B refer to points A and B of Figure 8.1.

For column C8 of Figure 8.2(a), whose section is a $10\times10\times72$ UC, for instance, the variation of the maximum moment in the member with the load factor is given by the full line $OABE$ in Figure 8.2(b). The moment reaches the reduced plastic hinge moment of the section at B when the applied load factor is just below the permissible value of 1·4 for λ_1. Equation (8.1) suggests that the required reduced plastic moment should be 330 kNm. From equation (8.2) the axial load in this member at $\lambda_1 = 1·4$ is 852 kN. Under this load, the lightest section that has a reduced M_p^1 of more than 330 kNm. in. is $12\times12\times79$ UC. with a nominal M_p of 460 kNm and M_p^1 of 398 kNm. Graph OA^1B^1 of Figure 8.2 is the result of a subsequent elastic-plastic analysis of the same four-storey frame with the newly selected section for column C8 replacing the old one. The analysis reveals that at $\lambda = 1·4$ the axial load in the column is 846 kN and M_p^1 of the selected section is 398 kNm as expected. The member finally developed a plastic hinge at a load factor of 1·46, as shown by point B^1.

In Figure 8.2 point B is very close to point D, and therefore it appears that a small strengthening of the section would prevent the formation of the plastic hinge in the member, thus satisfying the

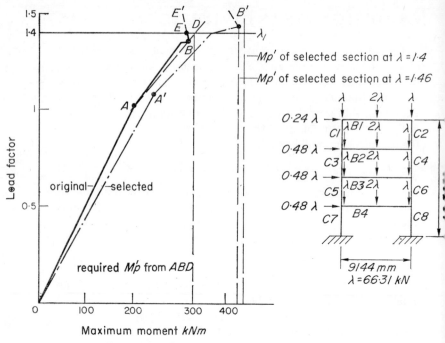

Figure 8.2. Behaviour of column C8 of a four-storey frame

design criterion prescribed earlier in this section. The method of design, however, has been found applicable to other members that develop hinges below a given load factor. Curve (*a*), which is drawn to a smaller scale in Figure 8.1, for instance, is for a member of an irregular frame that is designed later in this chapter. This member develops a hinge at a load factor of 1·2, shown by point *B*, which is considerably below the load factor of 1·4. Nevertheless the member was redesigned according to this method, and the result of a subsequent analysis of the structure, with the new section, is that shown by curve (*b*) of Figure 8.1. The member is still elastic at the load factor of 1·4.

A second example of design is that of a member which remains elastic in a frame up to collapse, but where collapse takes place below λ_1 (curve 2, Figure 8.1). For this member, the elastic-plastic analysis of the frame produces information about the variation of the maximum moment in the member up to the collapse state shown by points *B* and *F*. The extrapolation procedure used for the first design example can again be used to estimate the maximum

240

elastic moment in this member at the permissible load factor λ_1. This is represented by point D and the new section should have $M_p^1 > M_D$ at $\lambda = \lambda_1$.

A strong member in a frame is represented by Curve 5. This member develops plasticity at a load factor higher than that desirable for safety requirements. The section of this member may be reduced to $M_p^1 > M_D$, where M_D is the maximum moment in the member, as shown by point D, at $\lambda = \lambda_1$. This can be predicted accurately by interpolation between points B and C. On the other hand, Curve 1 represents a strong member that remains elastic even after point B above λ_1. This member can also be redesigned by interpolation between C and B and its section reduced so that M_p^1 is just larger than M_D given in the figure. In the case of the frame of Figure 8.2(a), for instance, an elastic-plastic analysis showed that at $\lambda = 1\cdot403$, the maximum moment in column C6 was 1216 kNm with the section being a $10\times10\times72$ UC. The member was therefore redesigned and the section reduced to a $10\times10\times60$ UC. A subsequent analysis proved that no further alteration could be made to the member without violating the design criterion.

For a member that follows Curve 3, it is necessary to ascertain that M_p^1 is not less than the maximum moment in the member as given by point D. In the subsequent analysis of the frame with some or all the other members altered, this member may change its behaviour into one of those given by the other curves. This, of course, changes the design procedure for the member accordingly.

In all the above cases, with the exception of Curve 3 the information required is the axial loads and maximum moments corresponding to points A, B and C of Figure 8.1. Equations (8.1) and (8.2) are then used to select sections with $M_p^1 > M_D$ at $\lambda = \lambda_1$. This procedure for the design of individual members can be incorporated in the elastic-plastic design of sway frames. This is done by first selecting a set of initial sections for the frame and carrying out elastic-plastic analyses of the frame under various loadings. From the results of these analyses the individual members can be redesigned to satisfy a set of design criteria and repeating the process of analyses would gradually remove the inaccuracy in selecting the initial member sections.

In every frame the elastic-plastic analysis reveals that, while some of the initially selected sections are underdesigned, such as those represented by Curve 4, others are invariably overdesigned and require reduction of their sections. Members represented by Curves 1 and 5 fall within this category and their sections could possibly be reduced. Redesigning the members therefore leads to a better distribution of material throughout the frame.

8.4. DESIGN CRITERIA

A sway frame is defined as one that resists deflections in its own plane through the bending stiffness of its members, i.e. a frame that has no other safeguard to limit these deflections. Because of this, the combined vertical and wind load action is the determining factor which causes failure by overall sway instability. It is assumed that the shape factor of the sections used is unity. For Universal Beams and Columns this assumption is reasonable and any slight inaccuracy is compensated by the strain hardening property of the material. It is also assumed that adequate safeguards, such as bracing and composite action, are provided, to prevent lateral instability out of the plane of bending.

A frame is considered satisfactory if elastic-plastic analyses under proportional loading reveal that none of the following design criteria are violated

(a) Under combined dead load, super load and wind load from either side the frame should not collapse below the permissible load factor λ_1.

(b) Under dead load and vertical superload, the frame should not collapse below the permissible load factor λ_2.

(c) No plastic hinge should develop in a beam below the load factor of unity and the frame should be entirely elastic under the working load.

(d) No plastic hinge should develop in a column below the permissible load factor λ_1 under combined loading, or λ_2 under vertical loading.

Usually the permissible load factors λ_1 and λ_2 are taken as 1·4 and 1·75 respectively, which correspond to the 1·25 stress ratio allowed in BS. 449. In the examples given later, these values have been adopted. However, any other values that comply with current codes can be used. Strictly speaking the last two design criteria are not required by BS. 449. However, it is considered inadvisable to allow hinges to develop in a frame under the working loads. The prevention of hinges in the columns below the permissible collapse loads, as criterion (d) imposes, reduces the possibility of local column instability due to torsional buckling. Furthermore, the formation of a hinge in a column reduces the overall stiffness of a frame owing to the excessive deflections that result, and therefore is detrimental to the safety of the frame.

On the other hand beams can, according to criterion (c), develop hinges as soon as λ exceeds the working load factor λ_W. In the case of beams, therefore, the design method for individual members described in the last section, can be used with point D, Figure 8.1, corresponding to λ_W instead of λ_1 or λ_2. As pointed out in Section 8.1, the design criteria are often selected to suit the design approach. This is the case with criteria (c) and (d) that impose a definite load factor, such as λ_1 and λ_W, against the formation of hinges in the members. It is noticed that these load factors have to be defined before equations (8.1) and (8.2) can be used.

A computer programme, lacking human judgement, devised to carry out an automatic design of frames may select, during the process of satisfying the suggested design criteria, a section for a particular column that is smaller than that for the column above it. For this reason, facilities should be provided in the actual programme to avoid this occurrence by making the section of the lower column not smaller than that of the column above it. In order to satisfy practical considerations, facilities should also be available to ensure, when required, that more than one member is designed out of the same section.

8.5. DESIGN PROCEDURE

The steps given here are basic ones outlining the main features of the design method. The actual procedure, however, includes a number of intermediate steps that help to reduce computer time. These are discussed in the Section 8.6.

(i) A set of lower-bound sections is selected for the members. To avoid collapse under vertical loading at a load factor λ_2, the lower-bound beam sections are selected so as to avoid collapse by a rigid-plastic beam type mechanism. For a beam under uniformly distributed loading, for instance the selected sections should have

$$M_p > \lambda_2 WL/16, \tag{8.3}$$

where W is the total load on the beam with span L. No reduction in such sections can be made without violating criterion (b). A lower bound section for a column is given by

$$A_c > \lambda_2 W_s/f_y \tag{8.4}$$

where f_y is the yield stress and A_c is the area of the selected section. W_s is the sum of half the loading acting on the adjacent beams and the load acting on the columns above, i.e. the static load. As a rule these lower-bound sections are much weaker than the final sections.

(ii) The frame is then analysed elastic-plastically, first with wind loads acting from one side and then from the other. From each analysis the maximum moments and axial loads corresponding to points A, B and C of Figure 8.1 are calculated. The maximum moments ever attained in members whose behaviour corresponds to that shown in Figure 8.1 by Curve 3 are also obtained, together with the corresponding axial loads.

(iii) From this information the members are redesigned to satisfy design criteria (c) and (d). In the case of a column the values of the maximum moment and axial loads corresponding to point D in Figure 8.1 at $\lambda = \lambda_1$ are calculated and used to select the new sections so that $M_D < M_p^1$. In order to avoid local column instability due to torsional buckling, the selected section is tested for this in a manner due to Horne[18]. According to criterion (c) a beam can acquire a hinge once the working loads are exceeded. For a beam, therefore, the lightest section that remains elastic at $\lambda = \lambda_W$ is selected. This is done in a similar manner to the design of columns except that when using equations (8.1) and (8.2), λ_W replaces λ_1.

(iv) At this stage two sections are available for any member, each one being selected from the results of an elastic-plastic analysis for combined loading with wind acting from one side. The larger of these two sections is adopted for each member.

(v) Steps (ii) to (iv) are repeated until design criteria (a), (c) and (d) are satisfied with combined loading.

(vi) The frame is then analysed under vertical loading to check whether criterion (*b*) is satisfied, and final alterations are made if required. The procedure for the design of a member is similar to that for combined loading, except that columns are designed to remain elastic up to λ_2.

At this stage it is necessary to point out that the procedure outlined above is subject to two assumptions:

(a) That, in the case of sway frames, the effect of combined loading is uppermost. This is so for most frames and step (vi), therefore, only makes minor alterations to the actual design. For most frames therefore, the elastic-plastic analysis under vertical loading becomes useful only as a check that criterion (b) is actually satisfied. Nevertheless there are cases of sway frames that are weak under vertical loading, and a typical case is given later in this chapter as an example.

(b) Once hinges have developed in the columns, the overall stiffness of the frame is reduced considerably, resulting in the final collapse. It has been assumed therefore that, by preventing hinge formation in the columns before λ_1, the frame is made sufficiently stiff to withstand λ_1. Hence criterion (a) under combined loading is automatically satisfied. This is a valid assumption for most sway frames, where the loss in the overall frame stiffness is due to plasticity in the columns. However, it is possible for all the hinges to develop in the beams resulting in a premature collapse below λ_1 with elastic columns. This type of failure is due to both the effects of joint translation (i.e. $p\Delta$ effect) and loss of stiffness at the joints due to the fortion of hinges in the beams.

8.6. ECONOMY IN COMPUTER TIME

The design procedure presented in the previous section is an iterative process involving repeated elastic-plastic analyses of the frame. These require a substantial amount of computer time, because detection of a hinge requires repeated solution of the stiffness equations. The construction of these equations themselves requires the calculation of the stability functions of the members. Furthermore, every time a hinge is developed in the frame the stiffness matrix has to be modified. Details of analyses suitable for large frames with particular reference to economy in computation have been given in various sections of Chapter 6. A further substantial amount of computer time can be saved by the introduction of a number of intermediate steps in the design procedure. Some of these steps are as follows:

(i) Starting an analysis with an unrealistic set of lower-bound sections, such as those calculated by equation (8.4), can be wasteful of computer time. Instead, a set of more realistic initial trial sections can be read into the computer as data, or, alternatively, selected automatically by making use of the practical features of the frame. In the case of rectangular frames, for instance, an improved set of initial sections can be selected for the top external columns by applying the rigid plastic theory under vertical loading. The external columns below the top storey are then given larger sections to prevent reverse column taper. Internal columns may be given the same initial sections as the external ones. For non-rectangular frames, similar procedures can be used to initiate the first analysis with a more realistic set of sections. The set of lower-bound sections of equation (8.4) remains useful for avoiding the testing of totally unrealistic sections during the scanning process.

(ii) A second intermediate step incorporated in the computer programme is called the 'non-linear elastic design' step because it makes use of the non-linear elastic behaviour of the frame. Taking axial loads into consideration, the frame is analysed elastically under combined loading at a constant factor λ_1. The resulting axial load and maximum moment in each column are used to redesign the member so that it remains elastic at this load factor. For a beam, on the other hand, the maximum moment is factored by $1/\lambda_1$ to correspond to the working load factor $\lambda_W = 1$. The resulting moment is then used to redesign the beam. This intermediate step helps to bring points B and D of Figure 8.1 closer and thus fewer elastic-plastic analyses are required. In the case of unsymmetrical frames this step is repeated with wind acting in the opposite direction.

(iii) Figure 8.1 shows that the information required for a redesign becomes available once a hinge is located above λ_1 or λ_2. At this stage the analysis is therefore terminated. Thus, in many cases, the analysis is carried out for only a small portion of the total time required to analyse the frame up to collapse.

(iv) Finally, once one of the design criteria is violated, a re-design of the structure becomes necessary. To save computer time, therefore, the analysis is terminated as soon as this happens. The information so far available is then used to redesign the frame. From Figures 8.1 and 8.2 it can be seen that terminating

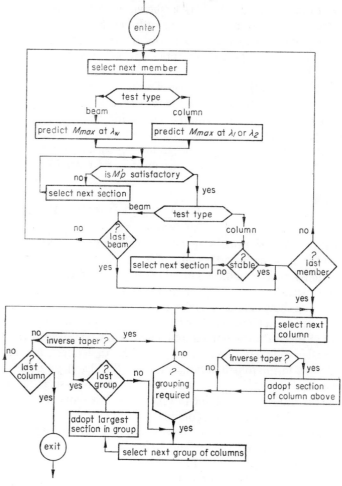

Figure 8.3. Flow diagram for design sub-routine

the analysis anywhere after the first hinge does not cause a sub-stantial loss of accuracy and, in most cases, entails no alteration in the choice of the required section. Furthermore, in the early stages of the design process, accuracy in selecting sections is not necessary. Therefore, at this stage, effort is directed to saving computer time and any inaccuracies introduced are rectified by the analyses in the later stages of the process.

A flow diagram showing the main steps of a subroutine (proce-dure) to redesign the members is shown in Figure 8.3. This sub-routine can be used in conjunction with the elastic-plastic analysis programme of Chapter 6 to carry out the entire design of frames automatically. In Figure 8.3 facilities are included to prevent re-verse column taper and also to group several members together so that they all have the same section. During the process of grouping the members together, it is possible to select a section for a group of columns which is smaller than the section for the group above. For this reason it is necessary to test and rectify this defect as shown in the flow diagram. A detailed programme procedure for an elastic-plastic design of sway frames is also given in reference 19.

8.7. DESIGN OF FOUR-STOREY FRAME

The four-storey frame is the one shown in Figure 8.2(a). This is a typical regular frame and because of its simplicity it is used here for an extensive investigation into the design of sway frames. The various design stages are given in Table 8.1. the lower-bound sec-tions being shown in the second column. As expected, the non-linear elastic design step changed the column sections drastically and the result of this is shown in the third column. Only one cycle of elastic-plastic design was required to select the final design sections. These are shown in the fourth column of Table 8.1. The load deflection curve, under combined loading, obtained by analysing the final design, is shown as a full line in Figure 8.4. The sequence of hinge formation, for vertical and combined loading is also shown in the figure. It may be seen that collapse takes place at a load factor of 1·49 under combined loading and 1·8 under vertical loading. Also, under combined loading the frame remains elastic up to a load factor of 1·08. No hinge develops in the columns below the permissible load factor for either loading case.

248

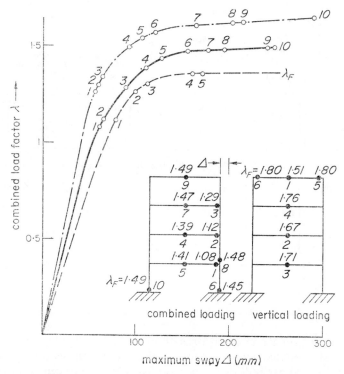

Figure 8.4. Four-storey frame: Behaviour of final design

Figure 8.5 shows the manner in which convergence to the final design is achieved starting from different initial sets of sections. The three designs were initiated with different sections weighing 56·1 kN, 47·6 kN and 33 kN. The final design was, therefore, approached from both sides and the three routes converged to the same set of sections. This indicates that the selection of the final design is independent of the initial trial sections or the route taken. It is noticed that the most economical route, as far as computer time is concerned is that starting from the lower-bound sections through the intermediate non-linear elastic design step, shown by the full line in the figure. Omitting the intermediate step and starting with a weight of 47·6 kN, which is quite near the final weight, necessitates a further elastic-plastic redesign. This is shown by the dotted line in Figure 8.5. Finally, starting with a heavy set of sections, with a weight of 56·2 kN, results in the slowest route to the final design.

Table 8.1 FOUR-STOREY FRAME

Member	Lower bound sections	Non-linear elastic sections	Final design	Design with top column heads pinned	Design restricted hinges in beams
B1 B2 B3 B4	$16\times7\times$ $\times40$ UB	$16\times7\times40$	$16\times7\times$ $\times40$	$21\times8\frac{1}{4}\times62$ $16\times7\times40$ $16\times7\times40$ $18\times7\frac{1}{2}\times45$	$16\times7\times40$ $16\times7\times40$ $18\times7\frac{1}{2}\times45$ $18\times7\frac{1}{2}\times45$
C1, C2		$10\times10\times60$	$10\times10\times$ $\times60$	$8\times8\times35$	$10\times10\times60$
C3, C4	$6\times6\times20$ UC	$10\times10\times60$	$10\times10\times$ $\times60$	$10\times10\times49$	$10\times10\times60$
C5, C6		$10\times10\times72$	$10\times10\times$ $\times60$	$10\times10\times72$	$10\times10\times60$
C7, C8		$10\times10\times72$	$12\times12\times$ $\times79$	$12\times12\times79$	$12\times12\times79$
Weight kN			48·9	48·0	50·0

A study of the reserved strength of this frame was carried out. This showed that it was possible to increase either the working wind loads by 20 per cent or the vertical loads by 7·6 per cent without reducing the load factor below the permissible value of 1.4. Thus, although the sway deflection of the frame is somewhat excessive, the frame itself has a reasonable reserve of strength over and above that required by BS. 449. The method is thus capable of selecting a set of economical sections that satisfy all the imposed design criteria. It is likely that, by removing one or more of these criteria, further saving in weight can be made. Apart from this, the method leaves the designer as the policy maker. For instance the heavy loading on the top floor of the frame leads to the selection of heavy columns. However, a more economical column section would have been obtained if, prior to the process, the designer had inserted real hinges at the column head of the top storey. This would have rendered the roof beam simply supported and reduced the column bending moments.

As far as the design method is concerned, such a decision would simply have required the alteration of the input data by allocating hinge numbers to the top columns. This frame was in fact redesigned

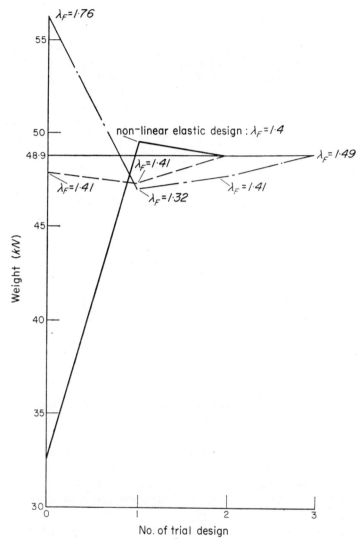

Figure 8.5. Four-storey frame progressions to final design

with these real hinges and the sections selected are shown in the fifth column of Table 8.1. As expected, the column sections are reduced while the beam sections are increased with a reduction of the weight of the frame from 48·9 kN to 48·0 kN i.e. about 2 per cent. In Figure 8.4 the load deflection curve of this frame is shown by the dotted lines, and it can be seen that the sway deflection of the frame has increased.

It was stated earlier that it is possible for a frame to collapse below λ_1 with the columns remaining elastic. As hinges develop in the beams the capacity of the latter to provide lateral stiffness to the columns is reduced, thus inducing overall instability of the frame. It

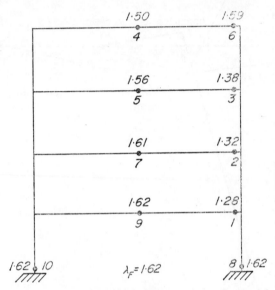

Figure 8.6. Formation of hinges in final design under combined loading with restriction of plasticity in beams

follows that an effective method of preventing this type of premature collapse is to limit the number of hinges allowable in the beams below a certain load factor. The introduction of such a restriction improves the carrying capacity of the frame and reduces its sway deflection.

In order to demonstrate these advantages, the same four-storey frame was designed with an extra criterion that prevents the formation of more than one hinge in any beam below the permissible load factor of 1.4 under combined loading. The sections selected are given in the last column of Table 8.1, where it is seen that the sections of the beams in the bottom two storeys are increased, leading to only a 2·5 per cent increase in the total weight. As a result of this, however, the carrying capacity of the frame went up from a load factor λ_F of 1·49 to 1·62—an increase of 8·7 per cent—while the sway deflection of the top floor at working load was reduced from 5·38 mm to 4·42 mm a reduction of 18 per cent. The load deflection curve of this frame is shown by the chain line in Figure 8.4, while the distribution of the plastic hinges is shown in Figure 8.6.

8.8. DESIGN OF IRREGULAR FRAMES

The efficiency of any automatic design method can best be tested against unfavourable design situations such as those encountered in the design of highly irregular frames, where the problem of sway is not only present but also difficult to assess. The irregularity of the shape of a frame, its loading and its unfavourable support conditions all aggravate the overall instability of the frame, necessitating an elastic-plastic type of approach.

In Figure 6.11 an irregular frame is shown in which instability due to excessive sway is very significant. This frame was designed by the elastic-plastic method of this chapter and it was found that the first elastic-plastic analysis of the trial sections put the failure load factor as low as 0·86. The sections selected by the redesign routine are those given in Table 6.1, the weight of the frame being 83·1 kN. The collapse load factors for this frame are 1·43 under combined loading and 2·81 under vertical loading. The positions of the hinges formed during a check elastic-plastic analysis are shown in Figure 6.11(b).

The highly irregular frame of Figure 8.7 displayed interesting behaviour during the various stages of the design process. The wind loading on the frame was calculated assuming exposure C of CP3, Chapter 5, and an effective height of 15·240 m. This gave a standard wind pressure of 670·32 N/m², but because of the irregular nature of the frame it was decided to apply this amount of wind pressure on both sides of the frame. The dimensions of the frame and the loading are shown in Figure 8.7.

After the non-linear elastic design step, two further cycles of iteration were required to obtain a safe design under combined loading. That is to say for each wind loading case, two elastic-plastic analyses were carried out. Each analysis confirmed that excessive axial loads in the columns and the inclined members reduced the overall frame stiffness and the plastic hinge moment of the sections drastically. The computer terminated each analysis

253

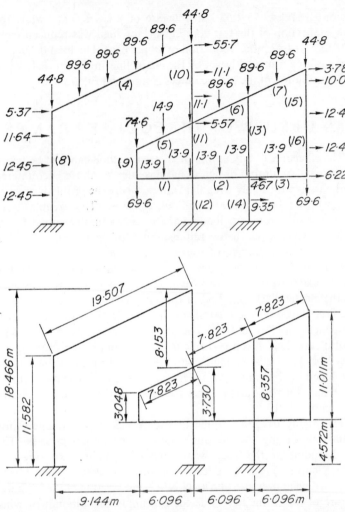

Figure 8.7. Irregular frame: dimensions, loading and member numbers

as soon as a design criterion was violated after the formation of two or three hinges only. A separate analysis was later carried out of the safe design, which showed that the frame developed seven hinges under each combined loading case and collapse took place at $\lambda_F = 1\cdot57$ for both wind directions.

In the case of sway frames, once a design is obtained which satisfies the combined loading case, the frame withstands the vertical loading. This is to be expected from sway frames where the overall sway instability is preponderant. Although excessive sway instability was also observed in the case of this frame, it was noticed that, once a safe design for combined loading had

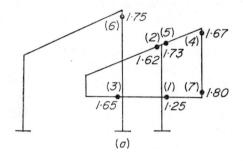

(a)

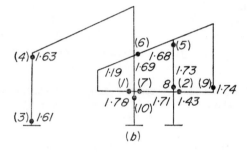

(b)

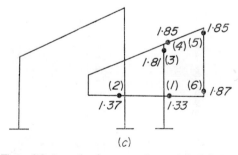

(c)

Figure 8.8. Irregular frame: analyses of final design
(a) Combined loading with wind from left $\lambda_F = 1.80$
(b) Combined loading with wind from right $\lambda_F = 1.78$
(c) Vertical loading $\lambda_F = 1.87$

255

been obtained, the frame proved to be unsatisfactory under vertical loading. When the computer proceeded with the analysis routine for vertical loading, the design criteria were violated at the formation of the third hinge. A subsequent redesign again proved unsatisfactory and this process went on for three cycles of iteration. Each time a design criterion was violated at an early stage of the analysis. When a satisfactory design was finally obtained, the frame collapsed at a load factor of 1·87 under vertical loading with seven hinges, at a load factor of 1·8 with wind from the left and 1·78 with wind from the right.

The selected sections are given in Table 8.2, while the distribution of plastic hinges in the frame is shown in Figure 8.8, where it can be seen that the hinges are distributed evenly throughout the frame. Indeed none of the members can be reduced any further. It was also noticed during the design process, that the weight of the frame increased gradually with every redesign and the operation was terminated as soon as the design criteria were all satisfied. This indicates that the final design is reasonably economical. The various analyses of the frame under vertical loading also revealed that, because of the irregular nature of the frame, considerable

Table 8.2 SECTIONS SELECTED FOR THE FINAL DESIGN.
(MEMBER NUMBERS ARE SHOWN IN BRACKETS IN
FIGURE 8.7).

Member	Section selected
1	24×9×68 U.B.
2	24×9×68 U.B.
3	27×10×84 U.B.
4	14×14·5×136 U.C.
5	14×14·5×103 U.C.
6	10×10×72 U.C.
7	14×14·5×136 U.C.
8	14×14·5×136 U.C.
9	12×12×92 U.C.
10	14×14·5×103 U.C.
11	14×14·5×103 U.C.
12	14×14·5×103 U.C.
13	14×14·5×103 U.C.
14	14×14·5×103 U.C.
15	10×10×72 U.C.
16	12×12×79 U.C.

256

sway deformation occurred and this contributed towards instability. The analysis of the final design under vertical loading shows that the left-hand column deflected by 29 mm at the working load. This indicates that the frame is of the sway type even under vertical loading.

This example reveals how intricate is the design of sway frames. The various stages of the design would have been impossible to forecast and any prejudgement of the number or position of hinges, the type of collapse mode or the critical loading case would have proved of little advantage.

8.9. CONCLUSION

The soundness of any design procedure can only be tested when it is applied to large and irregular frames. This is because these frames give rise to various new problems that seldom occur when dealing with simple frames. Most of these problems actually become apparent at a late stage of the design process and thus cause considerable difficulty. This is particularly the case when previous experience has led the designer, at an early stage, to adopt a certain decision which later proves to be the main cause of the problem. The method presented in this chapter, originated and developed by the author and later by Anderson and the author[19], appears to be able to cope with most of these problems. However, the examples show that a variety of difficulties may present themselves and the design of sway frames therefore cannot easily be reduced to a straightforward routine procedure.

The main source of economy in design comes from redesigning the frames not by restrictive assumptions about the mode of collapse or the position of hinges, but rather by consideration of the actual behaviour of the frame. This leads to the redistribution of surplus material from strong parts to members that develop hinges at an early stage of loading. Economy in design is also achieved by starting from a weak set of sections and increasing the total weight of the frame gradually up to the final weight, terminating the operation as soon as the design criteria are satisfied.

For various reasons, once the design criteria have been satisfied, the resulting frame has a reserve of strength. For instance, because a continuous set of sections is not available in the safe load tables, the selected sections are larger than those actually required. The lower-bound criterion for preventing a beam type mechanism also provides added strength to the frame since some of the beam sections selected are often stronger than those required by the criteria for combined loading, these criteria being the dominant factors in sway frames. Avoiding column hinge formation below the permissible collapse load factor also assists in strengthening the frame. Apart from safeguarding against local instability, it reduces the sway deflections and hence the overall frame instability, due to $P\Delta$ effect. This follows from observations that as soon as hinges are developed in the columns in an asymmetrical manner, the critical load of the frame deteriorates drastically and causes a sudden collapse. Further sources of strength are inherent in the design procedure. For instance, in the case of irregular and unsymmetrical frames there are two sections for each member to choose from, one from each wind loading case. The stronger of these is adopted and this gives the frame extra strength whichever way the wind load may act. The practical considerations, such as preventing reverse column taper and grouping the members together so that they have the same section, also increase the strength of the frame.

It should be emphasised, however, that the design of sway frames has not reached its final stage and considerable research is required before this is achieved. A particular difficulty is to control the sway deflections of the frame to those recommended in the codes of practice. The elastic-plastic method of this chapter results in frames that are strong enough to carry the loads safely but the deflections of these frames are frequently more than those recommended. This indicates that, in the case of sway frames, the deflection criteria dominate the strength criteria. It will be shown in Chapter 10 that producing a design with satisfactory deflections is a difficult task that requires further research.

The present design method produces a reasonably economical design but there is no evidence that this design is the most economical. It is evident that once deflection limitations are satisfied the weight of the frame will be more than that imposed by strength requirements alone. Nevertheless a new approach has recently been developed that makes use of linear and non-linear optimisation techniques. This selects the most economical design irrespective of the design requirements. Some aspects of this approach together with a brief introduction to optimisation techniques will be given in Chapters 9 and 10.

258

CHAPTER 9

Optimum design of structures using linear programming

9.1. INTRODUCTION

The analysis of a statically determinate structure is independent of the size of its members. In order to design such a structure which satisfies a given set of stress limitations, the member forces are first calculated using equations of static equilibrium. The cross-sectional properties of the members are then selected to satisfy the stress requirements. Provided that there is a continuous set of sections to choose from, the weight of the structure can be minimized without difficulty.

The analysis of hyperstatic structures, however, requires a knowledge of the member properties such as the area or the second moment of area of the sections. The design method for these structures is thus fundamentally different from that of statically determinate structures. It is common practice to perform repeated feasibility check analyses of the structure with trial sections. Such a process is not only laborious but also confuses the theory of design by avoiding it altogether and carrying out a number of analyses instead. This weakness of the existing design approach was pointed out by Pippard[20] as early as 1922. Nevertheless, practising design engineers have continued with this approach up to the present day.

The design of a structure is strictly speaking an economical problem involving decision making and planning. The fundamental nature of this planning process is far wider than the field of theory of structures and embraces practically every human activity. Very often the mathematical methods involved can be simplified, but an examination and understanding of the nature of the actual problem is more subtle. In the past twenty years or so there has been a remarkable growth of interest in a new branch of mathematics known as the optimization problem, often referred to as the programming problem. This problem can be considered as one concerned with the allocation of scarce resources, men, machines and raw materials for the manufacture of one or more products in such a way that these meet certain specifications, while at the same time some objective function such as profit, cost, weight etc. is maximised or minimised.

9.2. LINEAR PROGRAMMING

Let us clarify this new concept with the aid of a simple example. Consider a factory that produces two types of concrete blocks. A tonne of the first type requires 2 batches of gravel, 2 bags of cement and 4 batches of sand. A tonne of the second type of blocks requires 4 batches of gravel and 2 bags of cement but no sand. There are 20 batches of gravel, 12 bags of cement and 16 batches of sand available. It is required to find out how many tonnes of each type should be manufactured so that maximum profit may be achieved. It is assumed that the first type of blocks yields £2 profit per tonne and the second type yields £3.

The above information can be shown in the form of Table 9.1.

Table 9.1

Input	Technical factors		Total units available
	Block Type 1 (x_1)	Block Type 2 (x_2)	
Gravel	2	4	20
Cement	2	2	12
Sand	4	0	16
Profit	2	3	

Assume that x_1 tonnes of Type 1 and x_2 tonnes of Type 2 are manufactured. The variables x_1 and x_2 cannot take negative values and thus they have to satisfy the conditions

$$x_1 \geqslant 0$$
$$x_2 \geqslant 0$$ (9.1)

From the first row of Table 9.1 it is clear that x_1 tonnes of Type 1 blocks require $2x_1$ batches of gravel, while x_2 tonnes of Type 2 require $4x_2$ batches. The total number of these batches should not exceed the available 20. This information can be put in algebraic form as:

$$2x_1 + 4x_2 \leqslant 20$$

Similarly the next two rows of the table give

$$2x_1 + 2x_2 \leqslant 12$$
$$4x_1 \leqslant 16$$

Grouping these inequalities together, we have

$$\left. \begin{array}{l} 2x_1 + 4x_2 \leqslant 20 \\ 2x_1 + 2x_2 \leqslant 12 \\ 4x_1 \leqslant 16 \end{array} \right\}$$ (9.2)

The pair of variables x_1 and x_2 that satisfy constraints (9.1) and (9.2) are called feasible solutions, or feasible programs. In the context of this chapter it is preferable to call them feasible designs. The total profit Z is given by the last row of Table 9.1 as

$$Z = 2x_1 + 3x_2$$ (9.3)

or

$$2x_1 + 3x_2 - Z = 0$$ (9.3a)

This equation is often called the objective function and, among the various solutions, we are interested in the one that gives the optimum profit. It is noticed that the constraints (9.1) and (9.2) and equation (9.3) are all linear, i.e. no power, reciprocal or product of x_1 or x_2 appears in these expressions. This is the origin of the term linear programming.

261

To solve this problem, the inequalities (9.2) are first converted to equalities by introducing new so called slack variables y_1, y_2 and y_3. These represent the unused resources, thus

$$\left.\begin{array}{l} 2x_1+4x_2+y_1 = 20 \\ 2x_1+2x_2+y_2 = 12 \\ 4x_1+0x_2+y_3 = 16 \end{array}\right\} \tag{9.4}$$

As one solution to these equations it is noticed that if $x_1 = x_2 = 0$, i.e. before we produce any blocks then:

$$\left.\begin{array}{l} y_1 = 20 \\ y_2 = 12 \\ y_3 = 16 \end{array}\right\} \tag{9.5}$$

This is called the initial feasible solution and the profit for this is $Z = 0$. This solution can be organised in a table containing all the necessary information. This is called the simplex table, which is:

Table 9.2

x_1	x_2		
2	4	20	y_1
2	2	12	y_2
4	0	16	y_3
2	3	0	$-Z$

This table is essentially the same as Table 9.1 except that the names of the variables x_1, x_2, not in the solution appear at the top of the table. Those that are in the solution, y_1, y_2, y_3, are shown to one side and the value of the objective function multiplied by -1 is given below them. The profit of each item is given in the last row.

The initial solution is given in Table 9.2 in which the last two columns correspond to equations (9.5). This solution can be improved by producing at least one of the two types. Blocks of Type 2 give better profit than Type 1 and hence, as a trial, it is worth starting with this type. From the second row of Table 9.2, the amount of cement available limits the production to $12/2 = 6$ tonnes of blocks Type 2. However from the first row of the table the amount of gravel limits it to $20/4 = 5$ tonnes. Thus a maximum of 5 tonnes of block Type 2 can be produced. The important element 4, shown underlined, in the column and row corresponding to x_2 and y_1 is called the pivot. This indicates that x_2 and y_1 are to be interchanged. The new solution is $x_1 = 0$ and $x_2 = 5$. The amount of material used by this operation is thus; gravel $4 \times 5 = 20$ batches; cement: $2 \times 5 = 10$ bags and sand: $0 \times 5 = 0$. This leaves us with $y_1 = 20 - 20 = 0$ batches of gravel, $y_2 = 12 - 10 = 2$ bags of cement and $y_3 = 16 - 0 = 16$ batches of sand. The profit $Z = 2 \times 0 + 3 \times 5 = £15$.

Replacement of variable y_1 by x_2 is carried out using the first of equation (9.4) which gives

$$x_2 = 5 - 0 \cdot 5x_1 - 0 \cdot 25y_1 \qquad (9.6)$$

Substituting this in the other two of equations (9.4) and in (9.3a) gives

$$2x_1 + 2(5 - 0 \cdot 5x_1 - 0 \cdot 25y_1) + y_2 = 12$$
$$4x_1 + 0(5 - 0 \cdot 5x_1 - 0 \cdot 25y_1) + y_3 = 16 \qquad (9.7)$$
$$2x_1 + 3(5 - 0 \cdot 5x_1 - 0 \cdot 25y_1) - Z = 0$$

Equations (9.6) and (9.7) can be simplified to become

$$0 \cdot 5x_1 + 0 \cdot 25y_1 + x_2 = 5$$
$$x_1 - 0 \cdot 5y_1 + y_2 = 2$$
$$4x_1 + 0y_1 + y_3 = 16 \qquad (9.8)$$
$$0 \cdot 5x_1 - 0 \cdot 75y_1 - Z_1 = -15$$

The simplex table for these equations is

Table 9.3

x_1	y_1		
0·5	0·25	5	x_2
1	−0·50	2	y_2
4	0	16	y_3
0·5	−0·75	−15	$-Z$

The coefficients of x_1 in equations (9.8) appear in the first column of Tables 9.3 under x_1, similarly for y_1. The values of the variables x_2, y_2 and y_3 as well as the negative of the objective function are given in the third column. The first column of Table 9.3 gives us the following information. If we want to make one tonne of blocks Type 1, we have to reduce the production of Type 2 by 0·5 tonne. In this way we use one extra bag of cement and 4 batches of sand. The profit, in this case, would grow by £0·5. The second column of the table, under y_1, informs us that to save one batch of gravel, we have to reduce the production of x_2 by 0·25 tonne. This will also save 0·5 bag of cement but no sand. The profit will fall by £0·75. Thus the first column informs us that the profit can grow. The steps required to do this are

Step 1. Select a pivot that satisfies the three conditions

(i) The pivot is in a column whose last (profit) element is positive and the largest in the row corresponding to Z. In Table 9.2, for instance, both x_1 and x_2 columns can be selected for the pivot but in the last row $3 > 2$ and thus the column under x_2 contains the pivot.

(ii) The pivot must be non-zero and positive, otherwise the equations cannot be solved.

(iii) Divide the numbers in the last column by the corresponding numbers in the pivotal column and the pivot is given by the smallest positive result. Underline the pivot.

Step 2. If the pivot equals a, calculate $b = 1/a$ and replace the pivot by b in the new simplex table.

Step 3. Calculate the new elements in the pivotal row by multiplying the corresponding elements of the previous table by b.

Step 4. Calculate the new elements in the pivotal column by multiplying the corresponding elements of the previous table by b.

To avoid confusion Tables 9.2, 9.3 and the elements, calculated by steps (2), (3) and (4) of the new table are given first.

Table 9.4

x_1	x_2		
2	4	20	y_1
2	$\underline{2}$	12	y_2
4	0	16	y_3
2	3	0	$-Z$

x_1	y_1			
0·5	0·25	5	x_2	$5/0·5 = 10$
1	$-0·50$	2	y_2	$2/1 = 2$
4	0	16	y_3	$16/4 = 4$
0·5	$-0·75$	-15	$-Z$	∴ interchange x_1 with y_2 and pivot $= 1$ underlined

y_2	y_1			
$-0·5$			x_2	$0·5 \times -1 = = -0·5$
1	$-0·5$	2	x_1	$-0·5 \times 1 = -0·5 \times$ $2 \times 1 = 2$
-4			y_3	$4 \times 1 = -4$
$-0·5$			$-Z$	$0·5 \times -1 = -0·5$

Step 5. Calculate the rest of the table, one column at a time. In each incomplete column there is an element already calculated. For column j let this element be C_j. For column 2, for instance $C_2 = -0·5$. Calculate any other element d_{ik} from

$$d_{ik} = d_{ik}^1 - C_j \cdot e_{ip} \qquad (9.9)$$

where d_{ik}^1 is the old element in the previous table, and e_{ip} is the element in the previous table which is on the same row i as d_{ik}^1 and the pivotal column p.

In the example being considered, the missing elements of the new table are therefore calculated as follows

265

Element d_{12} on row 1 and column 2, under y_1 is:

$$d_{12} = 0.25 - (-0.5 \times 0.5) = 0.5$$

Similarly

$$d_{32} = 0 - (-0.5 \times 4) = 2$$
$$d_{42} = -0.75 - (-0.5 \times 0.5) = -0.5$$
$$C_3 = 2$$
$$d_{13} = 5 - (2 \times 0.5) = 4$$
$$d_{33} = 16 - (2 \times 4) = 8$$
$$d_{43} = -15 - (2 \times 0.5) = -16$$

The new table is as shown in Table 9.5.

Table 9.5

y_2	y_1		
-0.5	0.5	4	x_2
1	-0.5	2	x_1
-4	2	8	y_3
-0.5	-0.5	-16	$-Z$

Table 9.5. states that 4 tonnes of blocks type 2 should be produced and 2 tonnes of blocks type 1. There will be 8 batches of sand left over as surplus and the profit $Z = £16$. y_1 and y_2 are not in the solution and therefore all the gravel and cement is consumed. This programme or design is optimum since the elements in the last row corresponding to the profit are all negative and thus any alteration of the table to save gravel or cement will cause losses of £0.5 in each case. The optimum solution is always obtained when the elements of the last row are all negative.

The simplex method of linear programming that was explained when solving the above example is applicable in a general way to any number of variables and constraints. Further details of the method and proofs of its credibility can best be obtained from text books on linear programming[21, 22]. However, a few extra features of the method are given here so that its application to the design of structures can be presented.

Table 9.6

x_1	x_2	x_3	x_4		
2	1	4	0	100	y_1
1	5	0	1	200	y_2
0	4	1	2	200	y_3
1	0	1	$\underline{1}$	150	y_4
2	-1	-2	1	0	$-Z$

x_1	x_2	x_3		
2	1	4	100	y_1
1	3	-0.5	100	y_2
0	2	0.5	100	x_4
$\underline{1}$	-2	0.5	50	y_4
2	-3	-2.5	-100	$-Z$

x_2	x_3		
5	3	0	y_1
5	-1	50	y_2
2	0.5	100	x_4
-2	0.5	50	x_1
1	-3.5	-200	$-Z$

y_1	x_3		
0.2	0.6	0	x_2
-1	-4	50	y_2
-0.4	-0.7	100	x_4
-0.4	1.7	50	x_1
-0.2	-4.1	-200	$-Z$

9.3. EQUALITY CONSTRAINTS

In many linear programming problems some of the constraints are equalities and, because of this, no slack variables can be added to these equalities. For this reason, in order to obtain an initial feasible solution, artificial variables are added to the equalities and to begin with, the pivots are chosen on rows containing these equalities until they are all driven out of the solution. As an example consider the problem

$$
\begin{aligned}
2x_1 + x_2 + 4x_3 \quad &\leqslant 100 \\
x_1 + 5x_2 \quad + x_4 &\leqslant 200 \\
4x_2 + x_3 + 2x_4 &= 200 \\
x_1 \quad + x_3 + x_4 &= 150
\end{aligned}
\tag{9.10}
$$

maximise $\qquad Z = 2x_1 - x_2 - 2x_3 + x_4$

with $\qquad x_1, x_2, x_3, x_4 \geqslant 0$

In Table 9.6 this problem is solved. In the first section of the table, y_3 and y_4 are artificial variables. To eliminate these from the solution the first two pivots are selected without any reference to the elements on the Z row of the table. Instead they are selected on rows containing y_3 and y_4 respectively. This can be done provided that the pivot is positive. It is also noticed that as soon as an artificial variable leaves the solution, the column belonging to it is omitted. Once the artificial variables have left the solution the normal simplex method is performed in the usual manner. It will be noticed that the final solution can not be improved and therefore it is the optimum solution.

9.4. CONSTRAINTS WITH SURPLUS VARIABLES

So far the inequalities considered have been with $\leqslant$ signs and slack variables were added to these to convert them to equalities. Similarly, constraints with $\geqslant$ signs are converted to equalities by subtracting surplus variables from them. However, these surplus variables are not treated as slack variables but in the same way as the actual variables x. Thus they are also added to the objective function with zero coefficients. In this manner, the value of the objective function does not alter. Furthermore, artificial variables have to be added to all equality constraints that have surplus variables. For instance in the problem

$$2x_1+x_2+x_3 \leqslant 30$$
$$x_1+x_2 \quad = 10$$
$$x_1+x_2+x_3 \geqslant 8 \qquad (9.11)$$
$$x_1, x_2, x_3 \geqslant 0$$

maximise $\qquad Z = - x_1+2x_2+4x_3,$

a slack variable y_1 is added to the first constraint, an artificial variable y_2 is added to the second and a surplus variable $-x_4$ as well as an artificial variable y_3 are added to the third. The quantity $0x_4$ is also added to the objective function. The problem thus become

$$2x_1+x_2+x_3+y_1 = 30$$
$$x_1+x_2 \quad +y_2 = 10$$
$$x_1+x_2+x_3-x_4+y_3 = 8$$

maximise $\qquad Z = -x_1+2x_2+4x_3+0x_4$

In the first two sections of the simplex table the pivots are selected as positive numbers in the rows containing y_3 and y_2 so that these are eliminated. The reader can work through this problem to obtain the optimum solution which is: $x_1 = 0, x_2 = 10, x_3 = 20,$ $y_1 = 0$ and $x_4 = 22$ with $Z = 100$.

9.5. MINIMISATION PROBLEMS

These are treated exactly like maximisation problems except that the objective function is first multiplied by -1. This is because maximising the negative of an objective function is the same as minimising the function. As an example the problem

$$\left. \begin{array}{c} 4x_1-2x_2+x_3 \leqslant -8 \\ x_1+x_2+2x_3 \leqslant 12 \\ x_1, x_2, x_3 \geqslant 0 \end{array} \right\} \qquad (9.12)$$

minimise $\qquad Z = -2x_1+2x_2-3x_3,$

is first altered to become

$$-4x_1+2x_2-x_3 \geqslant 8$$
$$x_1+x_2+2x_3 \leqslant 12 \qquad\qquad (9.13)$$
$$x_1, x_2, x_3 \geqslant 0$$

maximise $\qquad\qquad Z = 2x_1-2x_2+3x_3.$

It is noticed that the right-hand side of the first of constraints (9.12) is -8. Whenever a negative number appears on the right-hand side the constraint is multiplied by -1 throughout and its inequality sign is reversed. This is done in the first of constraints (9.13). Before constructing the simplex table, a surplus variable x_4 is subtracted from the first constraint and an artificial variable is added to it. The final problem thus becomes

$$-4x_1+2x_2-x_3-x_4+y_1 = 8$$
$$x_1+x_2+2x_3+y_2 = 12$$
$$x_1, x_2, x_3 \geqslant 0$$

m aximise $\qquad\qquad Z = 2x_1-2x_2+3x_3+0x_4$

9.6. MINIMUM WEIGHT DESIGN OF STRUCTURES

The optimisation techniques referred to in the previous sections can be utilised to design structures that are subject to a number of structural limitations and in which the weight or the cost of the structure is to be minimised. A structure must satisfy three sets of basic design requirements. These are the strength requirements, that is to say when the structure is in equilibrium it should be sufficiently strong to carry the applied loads safely. The deflection requirements, that is the structure should nowhere deflect more than the amount acceptable by the standard specifications or any other criteria. Finally, there are the compatibility requirements i.e., the various parts of the structure remain attached to each other after the loads are applied.

The manner in which these requirements are satisfied often depends on the structural theory employed to express these requirements. In the case of the simple plastic theory, for instance, it is

270

assumed that, at collapse, the frame develops a mechanism and hence becomes statically determinate. This renders the compatibility requirements superfluous. The vector L_r of the redundant forces become a null vector and therefore the matrices B_r and B_r^1 of equations (2.67) also became null matrices. These make the compatibility equations (2.63) vanish. In equation (2.62) the product $F_{br}L_J$ also becomes a null vector. Furthermore, since the rigid plastic theory neglects the deflections of a frame, it follows that the rest of equation (2.62) also becomes redundant as there remains no deflection requirement to be satisfied. Accordingly, the strength requirements form the only constraints of the optimisation problem and these can be constructed from the equations of equilibrium.

Before entering into the details of the optimisation problem, there are a number of assumptions that are made in conjunction with the minimum weight design of structures. First of all, it should be pointed out that it is not assumed that the lightest structure is always the cheapest. The lightest structure has good claim to be considered as the best possible design. However, the main reason that the minimum weight design is put forward is because, for the first time, it approaches the problem of design in a fundamental manner. Nevertheless, the minimum weight design approach makes a number of assumptions. For instance despite the fact that in practice there is only a finite number of sections, it is assumed that an infinite number of sections are available to choose from. In using integer programming[22] this assumption is not required to be made, but integer programming is often lengthy and impractical. Secondly, it is assumed that there is a smooth curve relating the weight w per unit length of a member to some of its structural constants. In reality the weight of a member is linearly related to the cross-sectional area and an elastic minimum weight design correctly uses this for the design of pin jointed structures. For rigidly-jointed frames, the elastic minimum weight design relates the weight of the member to its second moment of area. On the other hand the rigid plastic minimum weight design relates the weight to the full plastic moment of the section. In reality there is no smooth curve connecting the weight per unit length of a member to its structural constants but such sections can be manufactured. At present for most sections it is reasonable to assume that

$$\left. \begin{array}{l} w = aM_p^b, \\ w = cI^d \end{array} \right\} \tag{9.14}$$

271

where a, b, c and d are constants. Using the plastic theory the values of 2·86 for a and 0·6 for b are commonly adopted. Equations (9.14) can be used to form the objective function for the total weight Z of the structure as

and
$$\left. \begin{aligned} Z &= \sum_{i=1}^{m} aM_{pi}^b L_i \\[2ex] Z &= \sum_{i=1}^{m} cI_i^d L_i \end{aligned} \right\} \qquad (9.15)$$

where L_i is the length of member i in a structure of m members. An objective function of this form renders the optimisation problem non-linear and methods given in the next chapter are needed to design the structure. To use linear programming it is necessary to replace equations (9.14) by linear relationships such as

$$\left. \begin{aligned} w &= e+fM_p, \\ w &= g+iI \end{aligned} \right\} \qquad (9.16)$$

where e, f, g and i are constants. The error involved in this linearization process is very small. This is because there will be no need to consider a wide range of sections in any particular problem. Using equations (9.16), the first of equation (9.15) becomes

$$Z = \Sigma(e+fM_p)L$$

i.e.
$$Z = e\Sigma L + f\Sigma M_p L$$

The quantity $e\Sigma L$ is constant and with the simple plastic minimum weight design the total weight of the frame is therefore minimised when

$$Z = \Sigma M_p L \qquad (9.17)$$

is minimized. Similarly, for rigidly jointed frames, the elastic minimum weight design uses

$$Z = \Sigma IL \qquad (9.18)$$

as the objective function.

272

9.7. THE RIGID PLASTIC MINIMUM WEIGHT DESIGN

This design method uses the simple plastic theory to select sections for the various members of a frame so that its weight is minimum and its strength requirements are satisfied. The strength constraints can be constructed using the equilibrium conditions which can easily be derived from the virtual work equations for the individual mechanisms.

In the design problem, it is required to select sections for the various members and, until this is done, the properties of the sections are unknown. For this reason the position of the plastic hinges and the type of the collapse mechanism cannot be determined in advance. One approach to the design problem is to consider all the possible collapse mechanisms with all the possible positions of the plastic hinges. With these mechanisms the boundary of the feasible solutions is defined and within this the solution that minimises the objective function is selected. This is called the mechanism method.

9.8. DESIGN OF A PORTAL BY MECHANISM METHOD

As an example consider the design of the simple portal of Figure 9.1. In this frame it is required to find the full plastic hinge moments M_{p1} of the two columns and M_{p2} of the beam so that the weight of the frame is minimum. There is a total of six mechanisms with which collapse may take place. If M_{p1} is larger than M_{p2}, the hinges at joints B and C will form in the beam and cases 1 to 3 show the different mechanisms that can develop at collapse. The work equations for these cases give

$$(1)\ 4M_{p2} = WL/2$$

$$(2)\ 2M_{p1}+4M_{p2} = 3WL/2$$

$$(3)\ 2M_{p1}+2M_{p2} = WL$$

Similarly if M_{p1} is less than M_{p2}, the work equations for the

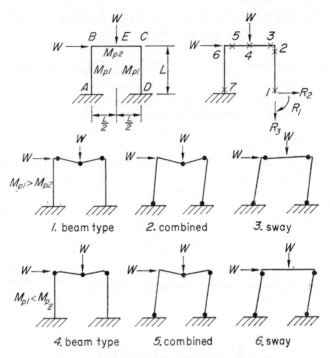

Figure 9.1. Possible mechanisms in a portal

mechanisms of cases 4 to 6 give:

$$(4)\ 2M_{p1}+2M_{p2} = WL/2$$
$$(5)\ 4M_{p1}+2M_{p2} = 3WL/2$$
$$(6)\ 4M_{p1} = WL$$

In case 1, the required value of M_{p2} is given by $WL/8$ and to prevent collapse with this type of mechanism the beam section should be selected with M_{p2} greater than or equal to $WL/8$

i.e. $M_{p2} \geqslant WL/8$

or $8M_{p2}/WL \geqslant 1$

Similarly to prevent collapse with the mechanism of case 2, we obtain

$$2M_{p1}/WL + 4M_{p2}/WL \geqslant 3/2$$

274

Denoting M_{p1}/WL by x_1 and M_{p2}/WL by x_2 the six mechanisms give the linear programming constraints as

$$8x_2 \geqslant 1$$
$$2x_1+4x_2 \geqslant 3/2$$
$$2x_1+2x_2 \geqslant 1 \qquad\qquad (9.19)$$
$$2x_1+2x_2 \geqslant 1/2$$
$$4x_1+2x_2 \geqslant 3/2$$
$$4x_1 \geqslant 1$$

The total length of both columns is $2L$ and the length of the beam is L. From equation (9.17) the objective function is therefore given as

$$Z = 2LM_{p1}+LM_{p2} \qquad\qquad (9.20)$$

which is to be minimised. Dividing both sides of this equation by WL^2 we obtain

$$Z^1 = Z/WL^2 = 2x_1+x_2 \qquad\qquad (9.20a)$$

and when Z^1 is minimum Z is minimum. Equation (9.20a) is there-fore used as the objective function for the constraints (9.19).

There are only two unknowns x_1 and x_2 to be determined and such a problem can easily be solved graphically. In Figure 9.2 x_1 and x_2 are chosen as the coordinate axes. The horizontal line AA^1 is the graph of equation $x_2 = 1/8$ and, for a design to be feasible, the first of constraints (9.19) restricts x_2 to values on or above this line. Similarly, the vertical line CC^1 is the graph of $x_1 = 1/4$ and the last of constraints (9.19) restricts x_1 to values on or to the right of this line. The two lines intersect at B. In order to satisfy the first and the last constraints (9.19) values of x_1 and x_2 have to be on or to the right of CB and on or above BA respectively. However, line DE is the graph for $2x_1+4x_2 = 3/2$ and a feasible design must, by virtue of the second of constraints (9.19), be on or above DE. The feasible region is shown shaded in the figure. Line HG is the graph for $2x_1+2x_2 = 1$. This line intersects the feasible region at one point D and hence there is only one feasible design in which the third of constraints (9.19) is active. Line JC^1 belongs to the fourth con straint. This line is outside the feasible region and thus it is always inactive. Finally, the fifth constraint is only active at one point D where the graph of $4x_1+2x_2 = 3/2$ touches the feasible region.

The straight lines, marked Z_1, Z_2 etc. are graphs of $2x_1 + x_2 =$ = Constant. One of these lines, that passes through the feasible region is also nearest to the origin O, has the smallest value of Z^1 and hence Z. Accordingly line KL, marked Z_{min}, gives the minimum weight design. This line passes through a feasible point D that decides values of x_1 and x_2 as both being one-quarter. Any other

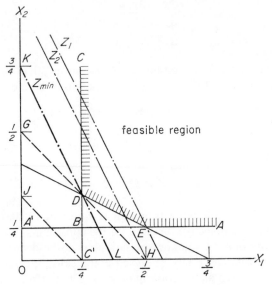

Figure 9.2. Graphical design of portal frame

line for equation (9.20a) nearer to the origin than KL avoids the feasible region and point D therefore gives the optimum design. The optimum full plastic moments of the sections are therefore

$$M_{p1} = x_1 WL = WL/4$$

and

$$M_{p2} = x_2 WL = WL/4$$

Equation (9.20) now gives the minimum value of the objective function as $3WL^2/4$.

9.9. MINIMUM WEIGHT DESIGN USING EQUILIBRIUM

Constraints (9.19) were obtained from the virtual work equations of all the possible mechanisms. In large frames with several members made out of various sections it is difficult to detect and consider all these mechanisms. For this reason the original equilibrium equations $P = BL$, given as equations (2.65), must be used to derive the constraints that govern the strength requirements. At simple plastic collapse a frame is statically determinate and thus the bending moments at every point are known. The construction of matrix B is therefore easily carried out.

Let us consider the same portal. This frame has three redundants R_1, R_2 and R_3 as shown in Figure 9.1 and there are seven points in the frame where a plastic hinge may develop. These are marked 1 to 7 in the figure and by taking moments about these points the bending moments are defined. Thus

$$
\begin{aligned}
M_1 &= R_1 \\
M_2 &= R_1 - LR_2 \\
M_3 &= R_1 - LR_2 \\
M_4 &= R_1 - LR_2 + LR_3/2 \\
M_5 &= R_1 - LR_2 + LR_3 + WL/2 \\
M_6 &= R_1 - LR_2 + LR_3 + WL/2 \\
M_7 &= R_1 \qquad\;\; + LR_3 + 3WL/2
\end{aligned}
\tag{9.21}
$$

The actual signs of M_1 to M_7 depend on the signs of R_1, R_2 and R_3 and these redundant forces can reverse their directions. In order to prevent the formation of a plastic hinge at a point, such as point 1, the section of the member must be selected so that M_{p1} is larger than M_1 irrespective of the sign of M_1. This means that

$$
M_{p1} \geqslant |M_1| \tag{9.22}
$$

The modulus sign of this constraint can be included in the linear programming problem by representing the constraint with two inequalities of the form

$$
\left.
\begin{aligned}
M_{p1} &\geqslant M_1 \\
\text{and} \qquad M_{p1} &\geqslant -M_1
\end{aligned}
\right\}
\tag{9.23}
$$

277

Substituting for M_1 from the first of equations (9.21) and rearranging

$$\left.\begin{array}{c} M_{p1} - R_1 \geqslant 0 \\ M_{p1} + R_1 \geqslant 0 \end{array}\right\} \tag{9.24}$$

Now the value of R_1 can be positive or negative and it is possible to dermit negative values to appear in the solution by replacing R_1 by ąwo variables R_1^1 and R_1^{11} thus

$$R_1 = R_1^1 - R_1^{11} \tag{9.25}$$

Substituting from (9.25) into (9.24) and multiplying both sides of the resulting inequalities by -1 so that their signs are altered from $\geqslant$ to $\leqslant$, we obtain

$$\begin{array}{c} R_1^1 - R_1^{11} - M_{p1} \leqslant 0, \\ R_1^1 + R_1^{11} - M_{p1} \leqslant 0 \end{array} \tag{9.26}$$

This procedure, equations (9.22) through (9.26), may be applied to all the other bending moments given by equations (9.21). Realising that, for the beam, M_{p2} should be larger than M_3, M_4 or M_5 the constraints of the linear programming problem become

$$
\begin{array}{lll}
R_1^1 - R_1^{11} & -M_{p1} & \leqslant 0 \\
-R_1^1 + R_1^{11} & -M_{p1} & \leqslant 0 \\
R_1^1 - R_1^{11} - LR_2^1 + LR_2^{11} & -M_{p1} & \leqslant 0 \\
-R_1^1 + R_1^{11} + LR_2^1 - LR_2^{11} & -M_{p1} & \leqslant 0 \\
R_1^1 - R_1^{11} - LR_2^1 + LR_2^{11} & -M_{p2} \leqslant 0 \\
-R_1^1 + R_1^{11} + LR_2^1 - LR_2^{11} & -M_{p2} \leqslant 0 \\
R_1^1 - R_1^{11} - LR_2^1 + LR_2^{11} + LR_3^1/2 - LR_3^{11}/2 & -M_{p2} \leqslant 0 \\
-R_1^1 + R_1^{11} + LR_2^1 - LR_2^{11} - LR_3^1/2 + LR_3^{11}/2 & -M_{p2} \leqslant 0 \\
R_1^1 - R_1^{11} - LR_2^1 + LR_2^{11} + LR_3^1 - LR_3^{11} & -M_{p2} \leqslant -WL/2 \\
-R_1^1 + R_1^{11} + LR_2^1 - LR_2^{11} - LR_3^1 + LR_3^{11} & -M_{p2} \leqslant WL/2 \\
R_1^1 - R_1^{11} - LR_2^1 + LR_2^{11} + LR_3^1 - LR_3^{11} & -M_{p1} & \leqslant -WL/2 \\
-R_1^1 + R_1^{11} + LR_2^1 - LR_2^{11} - LR_3^1 + LR_3^{11} & -M_{p1} & \leqslant WL/2 \\
R_1^1 - R_1^{11} + LR_3^1 - LR_3^{11} & -M_{p1} & \leqslant -3WL/2 \\
-R_1^1 + R_1^{11} - LR_3^1 + LR_3^{11} & -M_{p1} & \leqslant 3WL/2
\end{array} \tag{9.27}
$$

where R_2 is substituted by $R_2^1 - R_2^{11}$ and R_3 is substituted by $R_3^1 - R_3^{11}$.

278

The objective function is once again given by equation (9.20) in which M_{p1} and M_{p2} are the only two unknowns. The substitution $R^1 - R^{11}$ for R does not therefore affect the objective function. However, it should be pointed out that if, in a general linear programming problem, a given variable x can take positive as well as negative values, then the substitution $x^1 - x^{11}$ for x should be made wherever x appears in the constraints and the objective function. In the final solution either x^1 or x^{11} would appear but never both.

The simplex method can now be used to obtain the optimum value of the objective function, subject to constraints (9.27). Before the simplex table is employed slack variables y_1 to y_{14} are added to these constraints and altogether there will be fourteen constraints with twenty-two unknowns. There are fourteen slack variables, two unknown section properties M_{p1} and M_{p2} and six variables, R^1 and R^{11}, for the three unknown redundants. It appears that, for this portal, the problem is much bigger when using the equilibrium equations instead of the mechanisms. However, the equilibrium method shows its superpriority when designing larger frames.

It has been demonstrated that the minimum weight design of frames by the simple plastic theory can be made into a linear programming problem by introducing a number of simplifying assumptions. However, if any other structural theory is used, or if stress and deflection limitations are taken into consideration, the programming problem becomes non-linear. The procedure for such problems will be presented in detail in the next chapter.

EXERCISES

1. In the continuous beam ABC in Figure 9.3, caculate the values $M_{p1} M_{p2}$ for minimum weight, (a) for $M_{p1} \neq M_{p2}$ and (b) for $M_{p1} = = M_{p2}$.

Answer. (a) $320\lambda WL/33$, $120\lambda WL/11$; (b) $120\lambda WL/11$

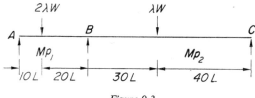

Figure 9.3

2. The three columns of the frame in Figure 9.4 are to be made out of the same section with fully plastic moment M_{p1}, while the beams are to be made out of a section with fully plastic moment M_{p2}. Calculate the value of M_{p1} and M_{p2} for minimum weight design using the plastic theory.

Answer. $M_{p1} = 8\lambda WL$, $M_{p2} = 52\lambda WL/3$

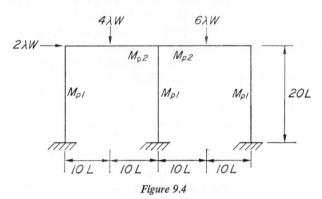

Figure 9.4

3. Using the plastic theory design the two storey frame in Figure 9.5 for minimum weight.

Answer. $M_{p1} = 50\lambda WL/3$, $M_{p2} = 80\lambda WL/3$, $M_{p3} = M_{p4} = 10\lambda WL$

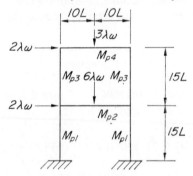

Figure 9.5

CHAPTER 10

Optimum design of structures using non-linear programming

10.1. INTRODUCTION

In Chapter 9 the general principles of linear programming and the simplex method of its solution were given. In order to apply linear programming to a given problem it is necessary to convert the constraints and the objective function to linear form. This was done in conjunction with the simple plastic minimum weight design where the constraints were linear but the objective function had to be linearised. Many problems, however, have non-linear constraints that are difficult or impossible to linearise. The minimum weight design of structures involving deflection or compatibility constraints, for instance, fall into this category. These problems are, by virtue of their nature, called non-linear programming problems.

Let us introduce the non-linear programming problem with the aid of a simple example. Consider the elastic design of the portal frame of Figure 2.9. for the case when all the members are of equal length $L = 1$ m and $E = 207$ kN/mm^2 throughout. The columns are to be made out of the same section. The frame is supporting a horizontal load H of 1 kN and it is required to select I_1 and I_2 so that the weight of the frame is minimum without allowing the sway deflection x_2 to exceed 4 mm. For this frame the joint equilibrium equations (2.49) become

$$\begin{bmatrix} H \\ 0 \end{bmatrix} = \begin{bmatrix} 2b_1 & -2d_1 \\ -2d_1 & 2(e_1+e_2+f_2) \end{bmatrix} \begin{bmatrix} x_2 \\ \theta_2 \end{bmatrix} \qquad (10.1)$$

Solving for x_2

$$x_2 = 0.5H/[b_1 - d_1^2/(e_1 + e_2 + f_2)]$$

Substituting for H, b_1, d_1, etc., we obtain

$$x_2 = 0.5 \times 10^9/\{2484I_1[1 - 3I_1/(4I, +6I_2)]\}$$

The problem requires that $x_2 \leqslant 4$ mm, the optimisation problem is thus to minimise

$$Z = 2I_1 + I_2 \tag{10.2}$$

subject to the contraints

$$0.5 \times 10^9/\{2484I_1[1 - 3I_1/(4I_1 + 6I_2)]\} \leqslant 4 \tag{10.3}$$

and

$$I_1 \geqslant 0, \quad I_2 \geqslant 0 \tag{10.4}$$

The objective function, which is linear, is derived using the linearised equation (9.18), but the first constraint is non-linear involving I^2 and $I_1 I_2$ as well as the reciprocals of I_1 and I_2. This problem is simple enough because it has only the two unknowns I_1 and I_2 and a graphical procedure can be devised to obtain the optimum design. To do this the boundary of the feasible region is first defined by equating the left-hand side of constraint (10.3) to 4 mm and plotting I_1 against I_2. Rearranging we obtain

$$I_2 = (9936I_1^2 - 2 \times 10^9 I_1)/(3 \times 10^9 - 59616I_1) \tag{10.5}$$

The graph of I_1 versus I_2 is shown in Figure 10.1 which is noticed to be non-lineal and convex towards the origin. The objective function, for constant values of Z, gives straight lines of which the line corresponding to minimum Z is shown tangent to the boundary of the feasible region. The point of tangency gives $I_1 = 77\,000$ mm^4, $I_2 = 60\,000$ mm^4 and $Z_{min} = 214$.

Although a matrix displacement method was used to begin with, the equations $L = KX$ were solved for the displacements before the constraints were derived. This implies that, in fact, matrix force method was utilized in the actual optimisation procedure. In general, any of the two matrix methods can be used to derive the structural constraints. However, the force method is superior in that it Formulates the programming problem with fewer constraints.

When analysing a structure the stiffness equations are conveniently solved for all the unknown displacements. These are then used to calculate the stresses in the members. Structural design, how-

282

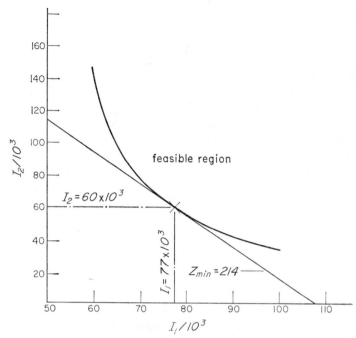

Figure 10.1. Graph of I_1 against I_2

ever, is the reverse operation of utilising the stress and displacement limitations to select the unknown sections. Generally, only some of the deflections are limited and the rest need not be involved in the process. It is because of this that matrix force method is more convenient. In this chapter, matrix force method is widely used to design statically determinate and indeterminate structures subject to stress and deflection limitations. Before proceeding with this, some basic principles of non-linear programming procedures, used in this chapter, are outlined. Further details on non-linear programming can be obtained from a textbook on the subject[23].

10.2. SOME PROPERTIES OF A CONVEX FUNCTION

Hitherto the design examples have always produced a unique set of sections for the various members. However, this may often not be the case. The problem may have several local optima of which only one is the global optimum. In minimisation problems, this case arises when the boundaries of the constraints are in the shape of valleys of variable depths. In this case it is possible to select one of the local optima. Indeed for the majority of non-linear programming problems the determination of the global optimum is a difficult or even an impossible task.

In minimisation problems, if the objective function and the set of constraints are convex then a local optimum is also global. In the last section, for instance, it was shown by Figure 10.1 that I_2, as given by equation (10.5), is a convex function of I_1 and it is noticed in Figure 10.1 that there is in fact a unique optimum solution to the problem. The properties of a convex function are

(i) In an n dimensional Euclidean space, the function $f(\chi)$ is convex if for any two values of χ_1 and χ_2 ($\chi_1 \neq \chi_2$) and for all values of λ, $0 \leqslant \lambda \leqslant 1$:

$$\lambda f(\chi_2) + (1-\lambda) f(\chi_1) \geqslant f[\lambda \chi_2 + (1-\lambda)\chi_1] \qquad (10.6)$$

Furthermore $f(\chi)$ is strictly convex for all λ, $0 < \lambda < 1$, if

$$\lambda f(\chi_2) + (1-\lambda) f(\chi_1) > f[\lambda \chi_2 + (1-\lambda)\chi_1] \qquad (10.7)$$

For instance $f(\chi) = 1/\chi$, for $\chi > 0$ is strictly convex because in the statement

$$\lambda/\chi_2 + (1-\lambda)/\chi_1 > 1/[\lambda \chi_2 + (1-\lambda)\chi_1], \qquad (10.8)$$

multiplying both sides by $\lambda \chi_2 + (1-\lambda)\chi_1$, which is always positive, we obtain

$$(1-\lambda)\, \lambda(\chi_1 - \chi_2)^2/\chi_1 \chi_2 > 0 \qquad (10.9)$$

Now the left-hand side of (10.9) is always positive and thus the statement (10.8) is true. Hence inequality (10.7) is satisfied for $f(\chi) = 1/\chi$ for all values of χ larger than zero.

(ii) The negative of a convex function $f(\chi)$ is the concave function $-f(\chi)$.

(iii) It follows from (ii) that a linear function is both convex and concave since its negative is also linear.

(iv) The sum of several convex functions is convex.

10.3. PIECEWISE LINEARISATION

A general non-linear programming problem requires the determination of n variables of χ_j which satisfy the constraints

$$\sum_{j=1}^{n} g_{ij}(\chi_j) \leqslant T_i \qquad (10.10)$$

with $j = 1, 1, n$; $i = 1, 1, m$

subject to the condition that

$$\chi_j \geqslant 0 \qquad (10.11)$$

and in addition, optimises the objective function

$$Z = \sum_{j=1}^{n} f_j(\chi_j) \qquad (10.12)$$

Some of the constraints (10.10) may also be equalities or inequalities with $\geqslant$ sign. The symbols g_{ij} and f_j refer to functions of χ_j. On the other hand the problem is linear provided that for each constraint

$$g_i(\chi_1, \ldots, \chi_j, \ldots, \chi_n) = \sum_{j=1}^{n} a_{ij}\chi_j \qquad (10.13)$$

and further provided that in the objective function (10.12)

$$f(\chi_1, \ldots, \chi_j, \ldots, \chi_n) = \sum_{j=1}^{n} c_j\chi_j \qquad (1.014)$$

where a_{ij} and c_j are known constants. For instance in constraints (9.11), $a_{11} = 2$, $a_{12} = 1$ and $a_{13} = 1$, while in the objective function $c_1 = -1$. $c_2 = 2$ and $c_3 = 4$. Linear problems that satisfy the limitations (10.13) and (10.14) can be solved by the simplex method.

It is possible to linearise problems, with separable functions, in a piecewise manner so that the simplex method may be used for the optimisation process. For instance, in Figure 10.2 the function $f(\chi)$ is divided between χ_0 and $\chi_\tau = \alpha$ into τ segments by $\tau + 1$ points. Here α is an upper bound on χ. The function $f(\chi)$ can be replaced by the piecewise linear function $f(\chi)$, shown dashed in the figure. Any (χ) where $\chi_k \leqslant \chi \leqslant \chi_{k+1}$ can be written as

$$\chi = (1-\lambda)\chi_k + \lambda x_{k+1} \qquad (10.15)$$

where

and

$$\left. \begin{array}{l} \lambda = (\chi - \chi_k)/(\chi_{k+1} - \chi_k) \\ 0 \leqslant \lambda \leqslant 1 \end{array} \right\} \qquad (10.16)$$

285

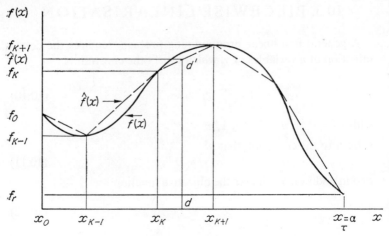

Figure 10.2. Piecewise linearisation of a curve

For example let $\chi_k = 5$, $\chi_{k+1} = 10$ and $\chi = 7$, then from (10.16)

$$\lambda = (7-5)/(10-5) = 2/5$$

and
$$1 - \lambda = 3/5$$

Equation (10.15) now gives

$$\chi = \tfrac{3}{5} . 5 + \tfrac{2}{5} . 10 = 7$$

as assumed.

For simplicity denote

and
$$\left. \begin{array}{l} \lambda_{k+1} = \lambda \\ \lambda_k = 1 - \lambda \end{array} \right\} \tag{10.17}$$

equation (10.15) becomes

$$\chi = \lambda_k \chi_k + \lambda_{k+1} \cdot \chi_{k+1}$$

and from Figure 10.2

$$\hat{f}(\chi) = dd' = \lambda_k f(\chi)_k + \lambda_{k+1} f(\chi)_{k+1} \tag{10.19}$$

More generally for any value of χ between zero and α it is possible to write

$$\chi = \lambda_0 \chi_0 + \lambda_1 \chi_1 + \ldots + \lambda_\tau \chi_\tau = \sum_{k=0}^{\tau} \lambda_k \chi_k \tag{10.20}$$

$$\hat{f}(\chi) = \lambda_0 f(\chi)_0 + \lambda_1 f(\chi)_1 + \ldots + \lambda_\tau f(\chi)_{\tau\alpha} = \sum_{k=0}^{\tau} \lambda_k f_k \tag{10.21}$$

286

and

$$\left.\begin{array}{l} \lambda_0+\lambda_1+\lambda_2+ \ \ldots \ +\lambda_\tau = \sum_{k=0}^{\tau} \lambda_k = 1 \\[2mm] \lambda_k \geqslant 0, \quad k = 0, 1, 2, 3, \ldots, \tau \end{array}\right\} \qquad (10.22)$$

This is provided that not more than two of the λ_k are positive and if $\lambda_k > 0$ and $\lambda_s > 0$ with $s > k$ then $s = k+1$. In this manner χ is uniquely determined and for any χ given by (10.20), $\hat{f}(\chi)$ is determined by (10.21) and will locate a point on the dashed curve of Figure 10.2.

As an example let $k = 4$, $k+1 = 5$, $\tau = 6$, $\chi_4 = 5$, $\chi_5 = 10$ and $\chi = 7$ which is between χ_4 and χ_5. Then from (10.16) and (10.17) it is found that

$$\lambda = \lambda_{k+1} = 2/5,$$
$$1 - \lambda = \lambda_k = 3/5$$

Equation 10.22 gives

$$\lambda_0+\lambda_1+\lambda_2+\lambda_3+3/5+2/5+\lambda_6 = 1$$
$$\therefore \ \lambda_0+\lambda_1+\lambda_2+\lambda_3+\lambda_6 = 0$$

Equation (10.20) therefore reduces to

$$\chi = \lambda_4\chi_4+\lambda_5\chi_5 = \tfrac{3}{5}\times 5+\tfrac{2}{5}\times 10 = 7$$

as assumed. Now if at χ_4 and χ_5 the ordinates $f(\chi)_4$ and $f(\chi)_5$ are 3 and 8 respectively then the ordinate $\hat{f}(\chi)$ at $\chi = 7$ is calculated from equation (10.21) as

$$\hat{f}(\chi) = \tfrac{3}{5}\times 3+\tfrac{2}{5}\times 8 = 5$$

Let us assume that in a given non-linear programming problem physical conditions impose an upper bound α_j on each χ_j so that $\chi_j \ngtr \alpha_j$ subdividing the distance from the origin to α_j into τ_j segments by τ_j+1 points with a typical point being χ_{kj}
and

$$\chi_{0j} = 0 \leqslant \chi_{1j} \leqslant \chi_{2j} \leqslant \ \ldots \ \leqslant \chi_{kj} \leqslant \chi_{k+1,j} \leqslant \ \ldots \ \leqslant \chi_{\tau_j}, \quad j = \alpha_j$$

then using equation (10.21)

$$\hat{g}_{ij}(\chi_j) = \sum_{k=0}^{\tau_j} \lambda_{kj} g_{ij}(\chi_k) \qquad (10.23)$$

and

$$\hat{f}_j(\chi_j) = \sum_{k=0}^{\tau_j} \lambda_{kj} f_j(\chi_k) \qquad (10.24)$$

287

where from (10.20)

$$\chi_j = \sum_{k=0}^{\tau_j} \lambda_{kj}\chi_{kj} \qquad (10.25)$$

and from (10.22):

$$\left. \begin{array}{l} \displaystyle\sum_{k=0}^{\tau_j} \lambda_{kj} = 1 \\ \text{with } \lambda_{kj} \geqslant 0 \text{ for every } k \text{ and } j \end{array} \right\} \qquad (10.26)$$

It should be stressed that for equations (10.23) and (10.24) the same subdivision should be used to divide the interval $0 \leqslant \chi_j \leqslant \alpha_j$ and χ_{kj} should be chosen so that each function $f_j(\chi_j)$ and $g_{ij}(\chi_j)$ is represented accurately. Using equations (10.23) through (10.26) the non linear programming problem given by (10.10), (10.11) and (10.12) is piecewise linearised to become

$$\sum_{j=1}^{n} \sum_{k=0}^{\tau_j} \lambda_{kj}g_{ij}(\chi_k) \leqslant T_i, \qquad (10.27)$$

$$\left. \begin{array}{l} \displaystyle\sum_{k=0}^{\tau_j} \lambda_{kj} = 1, \\ \lambda_{kj} \geqslant 0 \text{ for all } k \text{ and } j \end{array} \right\} \qquad (10.28)$$

and the objective function is

$$\hat{z} = \sum_{j=1}^{n} \sum_{k=0}^{\tau_j} \lambda_{kj}f_j(\chi_k) \qquad (10.29)$$

An approximate local optimum can be obtained for this problem by using the simplex method. It should be pointed out, however, that not more than two λ_{kj} are permitted to be positive for a given j and these two must be adjacent. It is also possible to obtain a global optimum for this problem if integer programming is employed.

10.4. STATICALLY DETERMINATE STRUCTURES

As an application of non-linear programming the minimum weight design of statically determinate pin jointed triangulated structures, is given here. The design uses the elastic theory and aims at satisfying stress as well as deflection requirements.

Without a knowledge of the cross-sectional areas, the member forces P can be calculated using equilibrium equations (2.57), thus

$$P = BL \qquad (10.30)$$

where L is the externally applied load vector and B is the load transformation matrix. For Q members the vector P is $\{P_1 \, P_2, \ldots P_j \ldots \ldots P_Q\}$ where j is a typical member and for m simultaneous external loads the vector L is $\{L_1 \, L_2 \ldots L_i \ldots L_m\}$. A design requirement is to prevent the stress in member j exceeding σ_j, it follows from Hooke's law that

$$P_j/A_j \leqslant \sigma_j \qquad (10.31)$$

where A_j is the unknown member area. The stress requirement is usually different from one member to another and the allowable stress vector σ is $\{\sigma_1 \, \sigma_2 \ldots \sigma_j \ldots \sigma_Q\}$. Using (10.30) and (10.31) the stress constraints for the structure become

$$aP = aBL \leqslant \sigma \qquad (10.32)$$

where the diagonal matrix a is

$$a = \begin{bmatrix} 1/A_1 & & & & 0 \\ & 1/A_2 & & & \\ & & \ddots & & \\ & & & 1/A_j & \\ & & & & \ddots \\ 0 & & & & 1/A_Q \end{bmatrix} \qquad (10.33)$$

For a single member constraints (10.32) impose that

$$D_j/A_j = [B_j]\{L\}/A_j \leqslant \sigma_j \qquad (10.34)$$

where D_j is an element obtained by multiplying row j of matrix B by the column vector L.

In a structure the deflection X under the applied loads are calculated from equations (2.60) which is

$$X = B'fBL \qquad (10.35)$$

where f is the member flexibility diagonal matrix and B' is the transpose of B. When the design requirements impose a set of upper bound deflections Δ on X, the deflection constraints become

$$B_b'fB_bL_b = B_b'fP_b \leqslant \Delta_b \qquad (10.36)$$

289

The suffix b is introduced as a reminder that a basic statically determinate structure is being considered. It should also be pointed out that not all the deflections are necessarily restricted. For this reason only the relevant rows of the square matrix $B'fB$ are considered. If a typical element of Δ restricts the deflection under a load L_i to δ_i, then constraints (10.36) impose that

$$\sum_{j=1}^{Q} a_{ij}/A_j \leqslant \delta_i \qquad (10.37)$$

The element on the left of this inequality is obtained by multiplying row i of the overall flexibility matrix $F = B'fB$ by the column of the applied loads. If p_{ij} is the force in member j when L_i has the value of unity and acts alone on the structure then

$$a_{ij}/A_j = P_j p_{ij} f_j$$

where f_j is the flexibility L/EA of member j. Constraint (10.37) therefore becomes

$$\sum_{j=1}^{Q} P_j p_{ij} f_j \leqslant \delta_i \qquad (10.38)$$

Structures are usually designed to sustain several load cases. In this chapter a given load case is referred to by the suffix h with a total of w load cases. It is also more economical to group several members together and manufacture them out of the same section. Each group is referred to by g and the total number of groups is referred to by G. Within a group, a member is referred to by j with the first member being J and the last member in the group being J'. The total length of each group is L_g and the area of the members in that group is A_g. For all the load cases the non-linear programming problem can now be formulated using constraints (10.34) and (10.37) as follows

Minimize

$$z = \sum_{g=1}^{G} L_g A_g \qquad (10.39)$$

subject to constraints

$$D_{jh}/A_g \leqslant \sigma_{jh} \qquad (10.40)$$

with $j = J, J+1$ up to J'; $g = 1, 1, G$ and $h = 1, 1, w$;

$$\sum_{g=1}^{G} \frac{1}{A_g} \sum_{j=J}^{J'} a_{ijh} \leqslant \delta_i \qquad (10.41)$$

with $i = 1, 1, m$; $h = 1, 1, w$, and

$$A_g \geqslant 0 \qquad (10.42)$$

The only unknowns ar the areas of the groups A_g. It is noticed that the objective function (10.39) minimises the actual volume of the structure and assumptions such as those given by equations (9.14) or (9.16) are not made. Furthermore the objective function is linear and therefore convex. The set of feasible solutions given by constraints (10.40) is, by virtue of inequality (10.8), with $\chi = A_g$, also convex.

However in constraint (10.41) the sign of each a_{ijh} depends on the sign of the product $P_{jh}P_{ij}$ of inequality (10.38), which can be either positive or negative. From the first two properties of a convex function it will be recalled that if a_{ij} is positive then a_{ij}/A_j is convex otherwise it will be concave. It follows that there is no definite proof that the set of feasible solutions given by constraint (10.41) is necessarily convex. A simple substitution of the form

$$x_g = 1/A_g \qquad (10.43)$$

suggested by Toakley[24], overcomes this difficulty. The non-linear programming problem thus becomes

Minimise
$$z = \sum_{g=1}^{G} Lg/x_g \qquad (10.44)$$

subject to constraints
$$D_{ih}x_g \leqslant \sigma_{jh}; \qquad (10.45)$$

$$\sum_{g=1}^{G} x_g \sum_{j=J}^{J'} a_{ijh} \leqslant \delta_i \qquad (10.46)$$

where
$$0 \leqslant x_g \leqslant u_g \qquad (10.47)$$

All the constraints are now linear and therefore convex. The non-linear objective function is also convex since Lg/x_g is always positive. This is because the total length Lg of the members of a group is always positive, while x_g is, by virtue of (10.47), non-negative and hence by inequality (10.8), the function $1/x_g$ is strictly convex. It follows that an optimum design to this problem is globally so.

Inequalities (10.47) impose an upper bound on x_g. From equation (10.43), $Ag = 1/x_g$ and when x_g reaches its upper bound u_g the quantity $1/x_g$ reduces to the lowest value $1/u_g$ that it can take. Hence $1/u_g = A_{g\,min}$ is the lower bound on the area of group g. To fix u_g therefore, $A_{g\,min}$ can be conveniently selected as the area of the smallest acceptable section for the group. That is to say

$$A_{g\,min} = |P_{jh}/\sigma_{jh}|_{min} \qquad (10.48)$$

This simplifies the programming problem by making constraints (10.45) superflous. Note that P_{jh} is the force in member j of group g under the external load case h which is calculated from statics and σ_{jh} is the known permissible stress in the same member when load case h is acting.

Inequalities (10.47) permit x_g to take the value of zero. This renders L_g/x_g in the objective function infinite. To prevent this it is necessary to make the substitution

$$x_{\omega} = y_g + l \qquad (10.49)$$

where y_g is a new non-negative variable and l is a constant lower bound on x_g so that when y_g becomes zero x_g will be equal to l. The value of l can be fixed by using equation (10.43) as

$$l = 1/A_{g\,max} \qquad (10.50)$$

$A_{g\,max}$ can conveniently be chosen as the area of the largest available section.

Making use of equations (10.48), (10.49) and (10.50) the non-linear programming problem becomes

Minimise
$$z = \sum_{g=1}^{G} L_g/(y_g + l) \qquad (10.51)$$

subject to

$$\sum_{g=1}^{G} y_g \sum_{j=J}^{J'} a_{ijh} \leqslant \delta_i - l \sum_{g=1}^{G} \sum_{j=J}^{J'} a_{ijh} \qquad (10.52)$$

$$0 \leqslant y_g \leqslant u_g - l \qquad (10.53)$$

Functions of variables y_g can be piecewise linearised using equations (10.20), (10.21) and (10.22). If the axis of a given y_g is split into t segments by $t+1$ points $f(y_g) = 1/(y_g + l)$ is piecewise linearised using equations (10.21) and the linearised function $\hat{f}(y_g)$ is then given by

$$\hat{f}(y) = \lambda_0 Y_0 + \lambda_1 Y_1 + \ \dots \ + \lambda_t Y_t = \sum_{k=0}^{t} \lambda_k Y_k \qquad (10.54)$$

where $Y_k = 1/(y_k + l)$

292

The suffix g is dropped for simplicity. Equation (10.20) also gives

$$y = \lambda_0 y_0 + \lambda_1 y_1 + \ldots + \lambda_t y_t = \sum_{k=0}^{t} \lambda_k y_k \qquad (10.55)$$

The programming problem is thus linearised. The linear programming problem is now

Minimise
$$z = \sum_{g=1}^{G} L_g \sum_{k=0}^{t} \lambda_{gk} Y_{gk} \qquad (10.56)$$

subject to

$$\sum_{g=1}^{G} \sum_{k=0}^{t} \lambda_{gk} y_{gk} \sum_{j=J}^{J'} a_{ijh} \leqslant \delta_i - l \sum_{g=1}^{G} \sum_{j=J}^{J'} a_{ijh} \qquad (10.57)$$

$$0 \leqslant \lambda_{gk} y_{gk} \leqslant u_g - l \qquad (10.58)$$

$$\left. \begin{array}{c} \sum_{k=0}^{t} \lambda_{gk} = 1 \\ \lambda_k \geqslant 0 \end{array} \right\} \qquad (10.59)$$

The design problem is now in a form that can be solved using the simplex method.

10.5. WORKED EXAMPLE

As a manually-worked example consider the pin-jointed frame of Figure 10.3. It is required to design this frame so that its weight is minimum while the vertical deflection at joint 2 is not allowed to

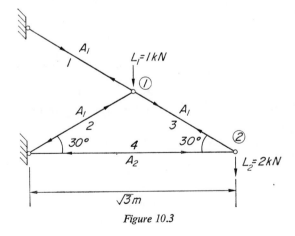

Figure 10.3

exceed 3 mm. The members and the joints are numbered in the figure. Members 1, 2 and 3 are to be made out of the same section with area A_1, while member 4 may be different and have area A_2. The length L of the members 1, 2 and 3 is 1 m. and their inclination is 60° to the vertical. Member 4 is therefore $\sqrt{3}$ m long. The modulus of elasticity E is 207 kn/mm² and the permissible stresses are $\sigma_T = = 0\cdot16$ kN/mm² in tension and $\sigma_c = 0\cdot1$ kN/mm² in compression. The largest available section has an area A_{max} of 100 mm².

Resolving horizontally and vertically at joints 1 and 2 we obtain

$$\begin{bmatrix} P_1 \\ P_2 \\ P_3 \\ P_4 \end{bmatrix} = \begin{bmatrix} 1 & 2 \\ -1 & 0 \\ 0 & 2 \\ 0 & -\sqrt{3} \end{bmatrix} \begin{bmatrix} L_1 \\ L_2 \end{bmatrix}, \quad \text{i.e.} \quad B = \begin{bmatrix} 1 & 2 \\ -1 & 0 \\ 0 & 2 \\ 0 & -\sqrt{3} \end{bmatrix}$$

(10.60)

If no restriction is imposed on the deflections of the frame and the areas are allowed to be all different then the above equilibrium equations give the areas of the members as 31·25, 10·00, 25·00 and 34·65 mm² respectively. However grouping the members together makes the area of the first three members A_1 as 31·25 mm² while member 4 has area A_2 of 34·65 mm².

The flexibility matrix f is

$$f = \frac{L}{E} \begin{bmatrix} 1/A_1 & & & 0 \\ & 1/A_1 & & \\ & & 1/A_1 & \\ 0 & & & \sqrt{3}/A_2 \end{bmatrix}$$

and

$$F = B'fB = \frac{L}{E} \begin{bmatrix} 2/A_1 & 2/A_1 \\ 2/A_1 & 8/A_1 + 3\sqrt{3}/A_2 \end{bmatrix}$$

From $X = FL$, the deflection d_2 at joint 2 is

$$d_2 = 2000(3/A_1 + \sqrt{3}/A_2)/69 \qquad (10.61)$$

with $A_1 = 31\cdot25$, $A_2 = 34\cdot65$, we find that $d_2 = 4\cdot23$ mm which is more than allowable.

294

There are two groups of members in the frame, i.e. $g = 1, 2$ and $G = 2$. It was shown that

$$A_{1\,min} = 10 \text{ mm}^2$$
$$\therefore u_1 = 1/10 = 0\cdot 1$$

and
$$A_{2\,min} = 34\cdot 65 \text{ mm}^2$$
$$\therefore u_2 = 1/34\cdot 65 = 0\cdot 0286$$

Further
$$l_1 = l_2 = 1/A_{max} = 1/100 = 0\cdot 01 = l$$

Thus $A_1 = 1/(y_1 + 0\cdot 01)$ and $A_2 = 1/(y_2 + 0\cdot 01)$

The total length for group $1 = L_{g1} = 3000$ mm, and the length of group $2 = L_{g2} = 1732$ mm. Using equations (10.51), constraints (10.52) or (10.61) and (10.53) the non-linear programming problem is

Minimise $\quad z = 3000/(y_1 + 0\cdot 01) + 1732/(y_2 + 0\cdot 01)$ $\qquad$ (10.62)

subject to

$$86\cdot 9y_1 + 50\cdot 2y_2 \leqslant 1\cdot 63, \qquad (10.63)$$
$$0 \leqslant y_1 \leqslant \cdot 090 \qquad (10.64)$$
$$0 \leqslant y_2 \leqslant 0\cdot 0186 \qquad (10.65)$$

Taking $t = 3$, the linearised objective function becomes

$$z = \sum_1^2 L_g \sum_{k=0}^3 \lambda_{gk} Y_{gk} = L_{g1} \sum_{k=0}^3 \lambda_{1k} Y_{1k} + L_{g2} \sum_{k=0}^3 \lambda_{2k} Y_{2k}$$

$$z = L_{g1}(\lambda_{1,0} Y_{1,0} + \lambda_{1,1} Y_{1,1} + \lambda_{1,2} Y_{1,2} + \lambda_{1,3} Y_{1,3})$$
$$\qquad + L_{g2}(\lambda_{2,0} Y_{2,0} + \lambda_{2,1} Y_{2,1} + \lambda_{2,2} Y_{2,2} + \lambda_{2,3} Y_{2,3})$$

The functions y_1 and y_2 are

$$y_1 = \lambda_{1,0} y_{1,0} + \lambda_{1,1} y_{1,1} + \lambda_{1,2} y_{1,2} + \lambda_{1,3} y_{1,3}$$
$$y_2 = \lambda_{2,0} y_{2,0} + \lambda_{2,1} y_{2,1} + \lambda_{2,2} y_{2,2} + \lambda_{2,3} y_{2,3}$$

Table 10.1 shows the values taken for y_1 and y_2. The corresponding values of Y_1 and Y_2 are calculated from: $Y_1 = 1/(y_1 + l)$ and $Y_2 = 1/(y_2 + l)$

Table 10.1

y_1	$y_{1,0}$	$y_{1,1}$	$y_{1,2}$	$y_{1,3}$
	0·00925	0·01100	0·01210	0·0140

y_2	$y_{2,0}$	$y_{2,1}$	$y_{2,2}$	$y_{2,3}$
	0·0090	0·0115	0·0136	0·0186

Thus

$$y_1 = 0.00925\lambda_{1,0} + 0.011\lambda_{1,1} + 0.0121\lambda_{1,2} + 0.014\lambda_{1,3}$$
$$y_2 = 0.009\lambda_{2,0} + 0.0115\lambda_{2,1} + 0.0136\lambda_{2,2} + 0.0186\lambda_{2,3}$$

The linearised programming problem is
Minimise

$$z = 3000\left(\frac{\lambda_{1,0}}{0.01925} + \frac{\lambda_{1,1}}{0.021} + \frac{\lambda_{1,2}}{0.0221} + \frac{\lambda_{1,3}}{0.024}\right)$$
$$+ 1732\left(\frac{\lambda_{2,0}}{0.019} + \frac{\lambda_{2,1}}{0.0215} + \frac{\lambda_{2,2}}{0.0236} + \frac{\lambda_{2,3}}{0.0286}\right)$$

subject to constraints

$$86.9(0.00925\lambda_{1,0} + 0.011\lambda_{1,1} + 0.0121\lambda_{1,2} + 0.014\lambda_{1,3})$$
$$+ 50.2(0.009\lambda_{2,0} + 0.0115\lambda_{2,1} + 0.0136\lambda_{2,2} + 0.0186\lambda_{2,3}) \leqslant 1.63;$$
$$\lambda_{1,0} + \lambda_{1,1} + \lambda_{1,2} + \lambda_{1,3} = 1;$$
$$\lambda_{2,0} + \lambda_{2,1} + \lambda_{2,2} + \lambda_{2,3} = 1;$$
$$\lambda \geqslant 0$$

For simplicity, substituting $X_1 = \lambda_{1,0}$, $X_2 = \lambda_{1,1}$, $X_3 = \lambda_{1,2}$, $X_4 = \lambda_{1,3}$, $X_5 = \lambda_{2,0}$ $X_6 = \lambda_{2,1}$, $X_7 = \lambda_{2,2}$, $X_8 = \lambda_{1,3}$ and taking v_1 as the slack variable and v_2 and v_3 as the artificial variables, completes the preparation for the simplex table. This is shown in Table 10.2 where the last row shows that the problem is optimum with $z = 21\,660$. The variables $X_3 = \lambda_{1,2} = 1$ and $X_6 = \lambda_{2,1} = 1$ are in the solution. These give

$$y_1 = 0.0121 \quad \lambda_{1,2} = 0.0121;$$
$$y_2 = 0.0115 \quad \lambda_{2,1} = 0.0115$$

Hence

$$A_1 = 1/(0.0121 + 0.01) = 45.2 \text{ mm}^2$$

and

$$A_2 = 1/(0.0115 + 0.01) = 46.5 \text{ mm}^2$$

In (10.61) these make the vertical deflection d_2 at joint 2 exactly 3·00 mm.

In Figure 10.4 the non-linear programming problem, given by equation (10.62) and constraints (10.63) to (10.65), is solved graphically. The feasible region is limited to that below the straight

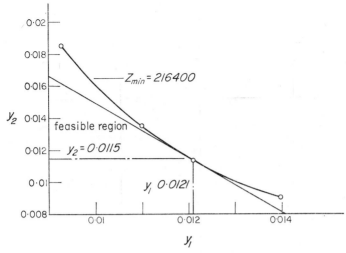

Figure 10.4. Graphical design of the pin jointed frame

line, which is obtained by using the equality sign of constraint (10.63).

The non-linear objective function for minimum z is shown with $z_{min} = 216\ 400$ as compared to $216\ 600$ obtained analytically.

10.6. STATICALLY INDETERMINATE STRUCTURES

The minimum weight elastic design of statically indeterminate pin-jointed structures form another example of non-linear programming. In order to design statically indeterminate structures the stress, the deflection as well as the compatibility constraints have to be satisfied. To satisfy the stress requirements in the members of a structure the member forces have to be calculated. This is done with the aid of the equilibrium equations (2.65) which is

$$P = BL = B_bL_b + B_rL_r \qquad (10.66)$$

These show that the member forces are dependent on the external loads L_b as well as the unknown redundant forces L_r in the structure. The values of L_r cannot be calculated without a knowledge of the cross-sectional properties which are also unknown.

297

Table 10.2

	X_1	X_2	X_3	X_4	X_5	X_6	X_7	X_8	
v_1	0.804	0.956	1.05	1.22	0.452	0.577	0.682	0.933	1.63
v_2	1	1	1	1	0	0	0	0	1.00
v_3	0	0	0	0	1	<u>1</u>	1	1	1.00
Z	-156 000	-143 000	-136 000	-125 000	-91 200	-80 600	-73 400	-60 600	0

	X_1	X_2	X_3	X_4	X_5	X_6	X_7	X_8	
v_1	0.804	0.956	1.05	1.22	-0.125		0.105	0.356	1.05
v_2	1	1	<u>1</u>	1	0		0	0	1.00
X_6	0	0	0	0	1		1	1	1.00
Z	-156 000	-143 000	-136 000	-125 000	-10 600		7200	20 000	80 600

Tableau 1

basis	X₁	X₂	X₄	X₅	X₇	X₈	
v_1	-0·246	-0·094	$\underline{0·17}$	-0·125	0·105	-0·356	0
X_3	1	1	1	0	0	0	1·00
X_6	0	0	0	1	1	1	1·00
Z	-20 000	-7000	11 000	-10 600	7200	20 000	216 600

Tableau 2

basis	X₁	X₂	v_1	X₅	X₇	X₈	
X_4	-1·45	-0·553	5·88	-0·735	0·618	2·09	0
X_3	2·45	1·553	-5·88	0·735	$\underline{-0·618}$	-2·09	1·00
X_6	0	0	0	1	1	1	1·00
Z	-4050	-917	-64 700	-2515	402	-2990	216 600

Tableau 3

basis	X₁	X₂	v_1	X₅	X₄	X₈	
X_7	-2·33	-0·896	9·50	-1·19	1·62	3·38	0
X_3					1		1·00
X_6					-1·619		1·00
Z	-3114	-557	-60 480	-2037	-650	-1632	216 600

299

The first term on the right-hand of equations (10.66) belongs to the statically determinate basic structure that is obtained by removing the redundant forces from the original structure. For N redundant forces the vector L_r appearing in the second term of the equation is $\{R_1 \ R_2 \ldots R_k \ldots R_N\}$ where R_k is a typical redundant force. The stress in a member is obtained by dividing the force in it by its cross-sectional area and for statically indeterminate structures the stress constraints are therefore given by

$$aP = aB_bL_b + aB_rL_r \leqslant \sigma \qquad (10.67)$$

where for Q members, including those that are chosen as redundants, σ is $\{\sigma_1 \ \sigma_2 \ldots \sigma_j \ldots \sigma_Q\}$ and a is the same diagonal matrix given by equation (10.33).

For a single member constraints (10.67) impose that

$$D_j/A_j + \sum_{k=1}^{N} e_{jk}R_k/A_j \leqslant \sigma_j \qquad (10.68)$$

where D_j is an element obtained by multiplying row j of matrix B_b by the column vector L_b of the external loads. On the other hand the constant e_{jk} is the element on row j and column k of matrix B_r. A_j is as before the area of member j. It is noticed that the first term on the left-hand side of inequality (10.68) is the same as the left-hand side of inequality (10.34) given for statically determinate structures.

The overall flexibility of a statically indeterminate structure is expressed by equations (2.62) and (2.63) These are

$$X_b = F_{bb}L_b + F_{br}L_r \qquad (10.69)$$
$$X_r = F_{rb}L_b + F_{rr}L_r \qquad (10.70)$$

X_b is the vector of displacements at points where the external loads L_b are acting. When the design requirements impose a set of upper bound deflections Δ_b on X_b, the deflection constraints become

$$F_{bb}L_b + F_{br}L_r \leqslant \Delta_b$$

from which upon using the definitions of equations (2.67) we obtain

$$B_b'fB_bL_b + B_b'fB_rL_r \leqslant \Delta_b \qquad (10.71)$$

Again the first term in this inequality is the same as that given by inequality (10.36) for statically determinate structures. The elements of matrix products $B_b'fB_bL_b$ and $B_b'fB_r$ each consist of a constant divided by the area of a member. If the deflections are restricted at m points to $X_b = \{\delta_1 \delta_2 \ldots \delta_i \ldots \delta_m\}$ and there are a total of Q members in a structure then each inequality in (10.71) corresponding to a given deflection δ_i becomes

$$\sum_{j=1}^{Q} \frac{a_{ij}}{A_j} + \sum_{k=1}^{N} R_k \sum_{j=1}^{Q} \frac{b_{ikj}}{A_j} \leqslant \delta_i \qquad (10.72)$$

where a_{ij}/A_j is obtained by multiplying the row i of matrix $F_{bb} = B_b' f B_b$ by the column vector of the applied loads L_b. Similarly b_{ikj}/A_j is obtained by multiplying row i of matrix $F_{br} = B_b' f B_r$ by the vector L_r of redundant forces.

During the process of satisfying the stress and deflection constraints, equations (10.70) also have to be satisfied since these express the compatibility conditions in the structure. These equations are therefore included in the optimisation process as compatibility constraints. The displacements Δ_r correspond to the redundants L_r and unless there is some lack of fit in the structure, Δ_r is a null vector. For each element Δ_k of the vector Δ_r a compatibility constraint takes the form of

$$\sum_{j=1}^{Q} \frac{c_{vj}}{A_j} + \sum_{k=1}^{N} R_k \sum_{j=1}^{Q} \frac{f_{vkj}}{A_j} = 0 \qquad (10.73)$$

with $v = 1, 1, N$, c_{vj} and f_{vkj} are the constants in the products $B_b' f B_b \, L_b$ and $B_r' f B_r$. It is noticed that the right-hand side of equations (10.73) is set to zero, since structures are usually designed to have no lack of fit.

For a total of w load cases applied to a structure whose members are grouped into G groups the non-linear programming problem can now be formulated as

Minimise
$$z = \sum_{g=1}^{G} L_g A_g \qquad (10.74)$$

subject to constraints

$$D_{jh}/A_g + \sum_{k=1}^{N} e_{jk} \times R_k / A_j \leqslant \sigma_j, \qquad (10.75)$$

with $j = J, 1, J^1$; $g = 1, 1, G$ and $h = 1, 1, w$;

$$\sum_{g=1}^{G} \frac{1}{A_g} \sum_{j=J}^{J'} a_{ijh} + \sum_{k=1}^{N} R_{kh} \sum_{g=1}^{G} \frac{1}{A_g} \sum_{j=J}^{J'} b_{ikj} \leqslant \delta_i \qquad (10.76)$$

301

with $i = 1, 1, m$ and $h = 1, 1, w$;

$$\sum_{g=1}^{G} \frac{1}{A_h} \sum_{j=J}^{J'} c_{vjh} + \sum_{k=1}^{N} R_{kh} \sum_{g=1}^{G} \frac{1}{A_g} \sum_{j=J}^{J'} f_{vkj} = 0 \qquad (10.77)$$

with $v = 1, 1, N$ and $h = 1, 1, w$;

$$A_g \geqslant 0 \qquad (10.78)$$

the symbols g, J, J' and h have the same meaning as those used in constraints (10.40) to (10.42) for statically determinate structures. The unknowns in this problem are the areas of the groups A_g and the redundant forces R_k and the programming problem is non-linear because it involves the reciprocal $1/A_g$ and the ratio R_{kh}/A_g. It will be shown that R_{kh} and A_g can be separated but because the compatibility constraints are equalities, the set of constraints produces a set of feasible solutions that are not necessarily convex. Thus a local optimum solution is not always globally so.

10.7. SEPARATION OF THE VARIABLES

In order to piecewise linearise the non-linear programming problem for statically indeterminate structures the variables have to be first separated. It is noticed that all the constraints have terms involving the ratio R_k/A_j (or R_{kh}/A_g when considering more than one loading case and grouping the members together). The two variables R_k and A_j in this ratio require separation and this can be done by substituting a variable x_{kj} for both R_k and A_j as

$$x_{kj} = R_k/A_j \qquad (10.79)$$

Taking the logarithms of both sides of this equation gives:

$$\log (x_{kj}) = \log R_k - \log A_j \qquad (10.80)$$

For each substitution of the type (10.79) it is thus necessary to include a constraint of the type (10.80) in the programming problem.

In an actual design, by virtue of inequality (10.78), the area of a section cannot take a negative value. Furthermore, reducing A_j to zero renders constraint (10.80) singular. To prevent both these defects a substitution of the form

$$\left. \begin{aligned} A_j &= y_j + l_j \\ y_j &\geqslant 0 \end{aligned} \right\} \qquad (10.81)$$

is necessary and imposes a constant lower bound l_j on A_j. Thus when the variable y_j is reduced to zero log A_j in (10.80) becomes log l_j.

In a structural problem the value of a redundant R_k can be positive, negative or zero. However, negative or zero values cannot be used for R_k in constraint (10.80). For this reason a substitution for R_k is also necessary. This may take the form of

$$\left.\begin{array}{c} R_k = r_k - c_k, \\ r_k \geqslant 0 \end{array}\right\} \qquad (10.82)$$

where r_k is a new variable. The constant $-c_k$ is the most negative value that the redundant R_k can take. This substitution is made into equation (10.80), where, when r_k is reduced to zero, log R_k becomes log $(-c_k)$. It is evident that a substitution of the type (10.82) is not sufficient to prevent the occurrence of logarithms of negative numbers. For this reason a further substitution is necessary for c_k, thus

$$c_k = U_k - d_k \qquad (10.83)$$

giving:
$$R_k = r_k - (U_k - d_k) \qquad (10.84)$$

Once the value of c_k is decided, a positive value is chosen for the constant d_k and equation (10.83) fixes the value of the constant U_k.

Similarly log (x_{kj}) may become singular and a constant B_{kj} must be added to the left-hand side of equation (10.79). However, before doing this the right-hand side of equations (10.81) and (10.84) can be substituted first in the original constraints. For instance constraint (10.72) becomes

$$\sum_{j=1}^{Q} \frac{a_{ij} - \sum\limits_{k=1}^{N} b_{ikj} \cdot U_k}{y_j + l_j} + \sum_{k=1}^{N} \sum_{j=1}^{Q} \frac{b_{ikj}(r_k + d_k)}{y_j + l_j} \leqslant \delta_i \qquad (10.85)$$

It is noticed that the second term of this equation involves $(r_k + d_k)/(y_j + l_j)$. Thus it is necessary to change equations (10.79) to become

$$\left.\begin{array}{c} x_{kj} + B_{kj} = (r_k + d_k)/(y_j + l_j), \\ x_{kj} \geqslant 0 \end{array}\right\} \qquad (10.86)$$

where B_{kj} is constant. Equations (10.86) alters the logarithmic constraint (10.80) into

$$\log (x_{kj} + B_{kj}) + \log (y_j + l_j) = \log (r_k + d_k) \qquad (10.87)$$

The value of B_{kj} can be chosen conveniently. Substituting from (10.86) into (10.85) and regrouping

$$\sum_{j=1}^{Q} \frac{H_{ij}}{y_j + l_j} + \sum_{k=1}^{N} \sum_{j=1}^{Q} b_{ikj}(x_{kj} + B_{kj}) \leqslant \delta_1 \qquad (10.88)$$

where

$$H_{ij} = a_{ij} - \sum_{k=1}^{N} b_{ikj}U_k \qquad (10.89)$$

constraints (10.68) and (10.73) or (10.75), (10.76) and (10.77) can be altered in exactly the same manner.

Upper and lower bounds can be established for $x_{kj} + B_{kj}$ as follows. From equations (10.86) it is noticed that $x_{kj} + B_{kj}$ is minimum when $r_k + d_k$ is a minimum, while $y_j + l_j$ is maximum. By virtue of (10.82) minimum r_k is zero; thus minimum $r_k + d_k$ is d_k. On the other hand $y_j + l_j$ is maximum when y_j is at its upper bound value $y_{j\,max}$. Hence the minimum value of $x_{kj} + B_{kj}$ is given by

$$(x_{kj} + B_{kj})_{min} = d_k / (y_{j\,max} + l_j) \qquad (10.90)$$

Further since both x_{kj} and B_{kj} are non-negative, because B_{kj} is selected to be so and x_{kj} is a variable and by virtue of (10.86) is non negative, it follows that if $x_{kj\,min}$ is zero then:

$$B_{kj} = d_k / (y_{j\,max} + l_j) \qquad (10.91)$$

On the other hand using equation (10.86) again, $x_{kj} + B_{kj}$ is maximum when $y_j + l_j$ is minimum and $r_k + d_k$ is maximum. That is to say when $y_j = 0$ and r_k takes its upper bound value $r_{k\,max}$, thus giving

$$(x_{kj} + B_{kj})_{max} = r_{k\,max} / l_j + d_k / l_j \qquad (10.92)$$

These bounds define the space of feasible solutions for the problem.

10.8. PIECEWISE LINEARISATION

The procedure of the last section separates the redundant variables R_k from the area A_j resulting in the replacement of these variables by x_{kj} and y_j. Functions of these variables can be piecewise linearised using equations (10.20), (10.21) and (10.22). If the axis of a given x_{kj} is split into s segments by $s+1$ points and the axis of a given y_j is split into t segments by $t+1$ points, then

304

$\hat{f}(x_{kj}) = x_{kj}+B_{kj}$ and $\hat{f}(y_j) = 1/(y_j+l_j)$ are piecewise linearised using equation (10.21). The linarised functions $\hat{f}(x_j)$ and $\hat{f}(y_j)$ are then given by

$$\hat{f}(x_{kj}) = \lambda_{kj0} \times X_{kj0} + \lambda_{kj1} \times X_{kj1} + \ldots + \lambda_{kjs} \times X_{kjs}$$

i.e.
$$\hat{f}(x_{kj}) = \sum_{K=0}^{s} \lambda_{kjK} \times X_{kjK} \qquad (10.93)$$

and

$$\hat{f}(y_j) = \sum_{K=0}^{t} \lambda_{jK} \times Y_{jK} \qquad (10.94)$$

where X_{kj} and Y_j are particular values of $f(x_{kj})$ and $f(y_j)$ respectively. For the case of a deflection constraint, for instance, substituting the right-hand side of equations (10.93) and (10.94) for $(x_{kj}+B_{kj})$ and $1/(y_j+1_j)$ into constraint (10.88) using (10.89), the deflection constraint becomes

$$\sum_{j=1}^{Q} H_{ij} \sum_{q=0}^{t} \lambda_{jq} \times Y_{jq} + \sum_{k=1}^{N} \sum_{j=1}^{Q} b_{ikj} \sum_{p=0}^{s} \lambda_{kjp} \times X_{kjp} \leqslant \delta_i \quad (10.95)$$

This procedure is used to linearize all the stress, deflection and compatibility constraints.

The logarithmic constraints (10.87) also have to be linearized using the same s and t segments for x_{kj} and y_j respectively. It is noticed from equation (10.87) that for any j

$$\log (x_{kj}+B_{kj}) + \log (y_j+l_j) = \log (x_{k1}+B_{k1}) + \log (y_1+l_1)$$
$$= \log (x_{k2}+B_{k2}) + \log (y_2+l_2) = \log (r_k+d_k) \quad (10.96)$$

In order to linearise these logarithmic constraints let

$$\log (x_{kj}+B_{kj}) = \sum_{p=0}^{s} \lambda_{kjp} . X_{jpk}' \qquad (10.97)$$

and

$$\log (y_j+l_j) = \sum_{q=0}^{t} \lambda_{jq} . Y_{jq}' \qquad (10.98)$$

where, it should be noticed, that λ in equations (10.97) and (10.98) are the same as those used in equations (10.93) and (10.94) respectively. But X' and Y' are different from X and Y. From equations

305

(10.97) and (10.98) using (10.96), the logarithmic constraints becomes

$$\sum_{p=0}^{s} \lambda_{k1p} \times X'_{k1p} + \sum_{q=0}^{t} \lambda_{1q} \times Y'_{1q} - \sum_{p=0}^{s} \lambda_{kjp} \times X'_{kjp} - \sum_{q=0}^{t} \lambda_{jq} \times Y'_{jq} = 0$$

(10.99)

Similar expressions are obtained for the other variables x and y. When the members of a structure are divided into G groups of equal cross-sections and there are N redundants in the structure there will be altogether $G(N-1)$ logarithmic constraints in the structure. The substitution $y+l$ for A must also be made in the objective function. In this case if Y''_{gq} is a particular value of $y_g + l_g$, the objective function becomes

$$z = \sum_{g=1}^{G} L_g \sum_{q=0}^{t} Y''_{gq} . \lambda_{gq}$$

(10.100)

In this manner the entire problem is prepared for use in conjunction with the simplex table.

10.9. DESIGN EXAMPLE

The simple structure of Figure 2.12(a) is selected to demonstrate the procedure for formulating the optimization problem. The structure is subject to a horizontal load H of 10 kN applied at the common joint where the horizontal displacement δ is not to exceed 0·254 mm. The dimension L is 2·54 m. The members are grouped together so that members 1 and 3 are both of area A_1 while members 2 and 4 are of area A_2. It is required that A_2 should not be less than 64·5 mm².

Selecting the forces P_2 and P_4 as the unknown redundants, equations (10.66) become

$$\begin{bmatrix} P_1 \\ P_2 \\ P_3 \\ P_4 \end{bmatrix} = \begin{bmatrix} 1/\sqrt{2} \\ 0 \\ -1/\sqrt{2} \\ 0 \end{bmatrix} . [H] + \begin{bmatrix} -1/\sqrt{2} & 0 \\ 1 & 0 \\ -1/\sqrt{2} & 1 \\ 0 & 1 \end{bmatrix} \begin{bmatrix} P_2 \\ P_4 \end{bmatrix}$$ (10.101)

The stress constraints are therefore given by

$$H/\sqrt{2}A_1 - P_2/\sqrt{2}A_1 \leqslant \sigma_1$$

$$P_2/A_2 \leqslant \sigma_2$$

and $\quad -H/\sqrt{2}A_1 - P_2/\sqrt{2}A_1 + P_4/A_1 \leqslant \sigma_3 \quad \Big\}$ (10.102)

$$P_4/A_2 \leqslant \sigma_4$$

where σ is the permissible stress in a member. The structural constraints (10.71) and (10.70) become

and $\quad \sqrt{2}H/A_1 + 0 \times P_2 - P_4/A_1 \leqslant E \times \delta/L$

$$0 \times H + (\sqrt{2}/A_1 + 1/A_2)P_2 - P_4/A_1 = 0 \quad \Big\}$$ (10.103)

$$-H/A_1 - P_2/A_1 + (\sqrt{2}/A_1 + \sqrt{2}/A_2)P_4 = 0$$

Finally the objective function z to be minimised is

$$z = 2\sqrt{2}LA_1 + (1 + \sqrt{2})LA_2$$ (10.104)

It is noticed that the objective function is linear. On the other hand the stress constraints have the same non-linearity of the form $1/A$ or P/A as the structural constraints. Because of the similariy of the stress and structural constraints the former are dropped for simplicity. This does not lead to any loss of generality in the problem which is now that of finding the minimum weight design that satisfies the deflection constraints only.

It can be shown by trial analyses that A_1 and A_2 are within the following ranges

$$5805 \leqslant A_1 \leqslant 7740 \text{ mm}^2,$$

$$64 \cdot 5 \leqslant A_2 \leqslant 1935 \text{ mm}^2.$$

Thus the minimum ratio A_2/A_1 is $64 \cdot 5/7740 = 1/120$ and the maximum ratio is $3/9$. Accordingly using the equality constraints (10.103), the values of the redundants for these two bounds become

For $A_2/A_1 = 1/120 : P_2 = 0 \cdot 00048$ kN and $P_4 = 0 \cdot 059$ kN.

For $A_2/A_1 = 3/9 : P_2 = 0 \cdot 4175$ kN and $P_4 = 1 \cdot 845$ kN.

Thus the upper and lower bounds of the redundant forces can be taken as

$$0 \leqslant P_2 \leqslant 0.5 \text{ kN.}$$

$$0 \leqslant P_4 \leqslant 2 \quad \text{kN.}$$

with two redundants $N = 2$ and k takes the values 1 and 2. The areas are divided into two groups thus $G = 2$ and $g = 1, 2$.

Selecting B_{11} arbitrarily as 0·25 say, equation (10.91) gives

$$d_1 = B_{11}A_{1\max} = 0·25 \times 7740 = 1935$$

Hence using (10.91) again for B_{12}:

$$B_{12} = d_1/A_{2\max} = 1$$

Similarly for $B_{21} = 0·25$, $d_2 = 0·25 \times 7740 = 1935$ and $B_{22} = 1$. Since the most negative values of P_2 and P_4 were taken to be zero, it follows using, equations (10.84) that

$$U_1 = U_2 = 1935$$

On the other hand the maximum value of P_2 is 0·5 and equation (10.84) gives $r_{1\max}$ as 0·5. With the maximum value of $P_4 = 2$, the same equation gives $r_{2\max}$ as 2. Hence

$$(r_1 + d_1)\max = 0·5 + 1935 = 1935·5$$

$$(r_2 + d_2)\max = 2 + 1935 = 1937$$

Using equation (10.92), it follows that; $(x_{11} + B_{11})_{\max} = 1935·5/$ $/5805 = 0·3334$. The values of other $(x_{kj} + B_{kj})$ are calculated in a similar manner. These are summarised in Table 10.3.

Table 10.3

Variable	Minimum Value	Maximum Value	Number of Segments
A_1	5805	7740	2
A_2	64·5	1935	22
$x_{11} + B_{11}$	0·25	0·3334	4
$x_{12} + B_{12}$	1	30·000	27
$x_{21} + B_{21}$	0·25	0·3337	6
$x_{22} + B_{22}$	1	30·031	29

Each variable can now be given a number of specific values within its range, thus dividing its non-linear variation into a number of linear segments. For instance A_1, being represented by two segments, see Table 10.3, can have the values 5805, 6774 and 7740. However, because the variables are separated using logarithms, a realistic representation is obtained by choosing the same number of segments between values of 1 and 10 on the one hand and 10 and 100 on the other. The number of segments into which each variable is divided is shown in Table 10.3.

The deflection and compatibility constraints are formulated as follows: In equation (10.94) for $j = 1$ and $K = 0$, with $A_1 = = (y_1+l_1)_0 = 5805$;

$$Y_{1,0} = 1/(y_1+l_1)_0 = 1/5805 = 0 \cdot 0001725$$

The second value of A_1 is 6774 and $Y_{1,1}$ is $1/6774$. Similarly $Y_{1,2}$ is $1/7740$. The first value of A_2 is $64 \cdot 5$ and hence

$$Y_{2,0} = 1/(y_2+l_2)_0 = 1/64 \cdot 5$$

In this manner all the values of Y_{jK} are calculated.

From constraints (10.103) $B'_b fB_b$ is $\sqrt{2}/A_1$ and L_b is 10, hence $a_{1,1} = 10\sqrt{2}$. The coefficient $b_{1,1,1}$ of P_2/A_1 in the first constraints (10.103) is zero, while the coefficient $b_{1,2,1}$ of P_4/A_1, is -1. Hence in constraint (10.88):

$$H_{1,1} = a_{1,1} - \sum_{k=1}^{2} b_{1,k,1} \times U_k = 10\sqrt{2} - 0 \times 1935 - (-1) \times 1935$$

$$\therefore H_{1,1} = 1949 \cdot 14$$

In the first of constraints (10.103), since A_2 does not appear in $B'_b fB_b$, the coefficient of $1/A_2$ is zero and therefore $a_{1,2} = 0$. The coefficients $b_{1,1,2}$ and $b_{1,2,2}$ of P_2/A_2 and P_4/A_4 are also zero, thus

$$H_{1,2} = a_{1,2} - \sum_{k=1}^{2} b_{1,k,2} \cdot U_k = 0$$

From these figures the coefficients of λ_{jq} in the first term of constraint (10.95) are prepared. The coefficient of $\lambda_{1,0}$ is $H_{1,1} \cdot Y_{1,0}$ which is equal to $1949.14/5805$, that of $\lambda_{1,1}$, is $H_{1,1} \cdot Y_{1,1}$ i.e. $1949.14/6774$ and that of $\lambda_{1,2} = m_{1,1} \cdot Y_{1,2} = 1949.14/7740$. Similarly for A_2 the coefficients of $\lambda_{2,0}$ up to $\lambda_{2,22}$ can be calculated. These are all zeroes.

In the second term of constraint (10.95) $b_{ikj} X_{kjp}$ is the coefficient of λ_{kjp}. This is prepared as follows: From the first of constraints (10.103) the coefficient $b_{1,1,1}$ of P_2/A_1 is zero. Thus in constraint (10.95)

Coefficient of $\lambda_{1,1,p} = b_{1,1,1} \cdot X_{1,1,p} = 0$, for $p = 0, 1, 2, \dots 5$.
Similarly

Coefficient of $\lambda_{1,2,p} = b_{1,1,2} \cdot X_{1,2,p} = 0$, for $p = 0, \dots 28$. However the coefficient $b_{1,2,1}$ of P_4/A_1 in the first of constraints (10.103) is -1. For $p = 0$ and $X_{2,1,0} = (x_{2,1}+B_{2,1})_0 = 0 \cdot 25$, see Table 10.3

309

Coefficient of $\lambda_{2,1,0} = b_{2,1,1}.X_{2,1,0} = -1.0\cdot25 = -0\cdot25$. The coefficients $b_{1,2,1}.X_{2,1,1}$, up to $b_{1,2,1}.X_{2,1,6}$ are calculated in the same manner making use of the point values of $x_{21}+B_{21}$. For instance:

$$b_{1,2,1}X_{2,1,6} = -1\times0\cdot3337 = -0\cdot3337$$

The value of $x_{21}+B_{21}$ is again taken from Table 10.3.

The coefficient $b_{1,2,2}$ of P_4/A_2 in the first of constraints (10.103) is zero and thus

$$b_{1,2,2} \sum_{p=0}^{29} \lambda_{2,2,p}X_{2,2,p} = 0.$$

This procedure is also used to calculate the coefficients of the other constraints (51).

Finally, for the logarithmic constraints (10.99), consider for instance $\Sigma\lambda 1q.Y'_{1q}$; these are

$$Y'_{1,0} = \log(y_1+l_1)_0 = \log 5805$$
$$Y'_{1,1} = \log(y_1+l_1)_1 = \log 6774$$
$$Y'_{1,2} = \log(y_1+l_1)_2 = \log 7740$$

The other terms of constraint (10.99) are prepared in a similar manner.

The final simplex table consists of eleven constraints, one deflection constraint, two compatibility, two logarithmic, one constraint of type (10.22) for each A_1 and A_2 and one of type (10.22) for each $x_{kj}+B_{kj}$. Altogether there are ninety-six structural (λ) variables, one slack variable for the deflection constraint and ten artificial variables for the equations. The table was reduced fifty-six times in order to obtain the first local optimum, giving A_1 as 6548 mm² and A_2 as 605 mm². Once a local optimum is obtained it is easy to proceed in search of other local optima. The fifth local required a total of eighty-eight reductions of the simplex table giving A_1 as 6729 mm² and A_2 as 167 mm². The value z of the objective function is thus $0\cdot494\times10^9$ mm³. Because there are only two unknown areas A_1 and A_2 to be found, the graphical method used for the example of the portal in the introduction to this chapter, can be used for this problem also. This gives A_1 as 6710 mm² and A_2 as 161 mm² and z as $0\cdot492\times10^9$ mm³. These two sets of results compare favourably.

310

10.10. DESIGN OF A RIGIDLY-JOINTED FRAME

The non-linear programming procedure used in the design of triangulated frames can be utilised for the minimum weight design of rigidly-jointed structures subject to deflection limitations only. Generally, in rigidly-jointed sway frames, the deflections are more critical than the strength requirements. Furthermore, methods such as that given in Chapter 8, are available for the elastic-plastic design of these frames to satisfy the strength requirements. It was pointed out in Chapter 8 that the elastic-plastic method of design yields good results in frames that are strong enough to carry the applied loads safely, but deflections of these frames are frequently more than those recommended. The deflection criteria of sway frames often dominate the strength criteria. The non-linear programming procedure therefore completes the design process of these frames by satisfying the deflection requirements.

For rigidly-jointed frames, stress constraints such as that given by equation (10.75) present difficulties because they have to be modified to include the second moment of area I of the sections as well as the distance Y of the extreme fibres from the neutral axes of the members. The relationship between I, Y and A, if such exist, are highly non-linear and to date no attempt has been made to include these in the stress constraints. Leaving out the stress constraints, the minimum weight design of rigidly-jointed frames, subject to deflection limitations only, is given by constraints (10.76) to (10.78) modified to include the second moment of area of the sections, thus

The deflection constraints are

$$\sum_{g=1}^{G} \frac{1}{I_g} \sum_{j=J}^{J'} a_{ijh} + \sum_{k=1}^{N} R_{kh} \sum_{g=1}^{G} \frac{1}{I_g} \sum_{j=J}^{J'} b_{ikj} \leqslant \delta_i \qquad (10.105)$$

The compatibility constraints are

$$\sum_{g=1}^{G} \frac{1}{I_g} \sum_{j=J}^{J'} C_{vjh} + \sum_{k=1}^{N} R_{kh} \sum_{g=1}^{G} \frac{1}{I_g} \sum_{j=J}^{J'} f_{vkj} = 0 \qquad (10.106)$$

Constraint (10.78) becomes:

$$I_g \geqslant 0 \qquad (10.107)$$

The objective function is also modified. This makes use of the

311

linearised form given by equation (9.18) and it is:

Minimise
$$z = \sum_{g=1}^{G} L_g I_g \qquad (10.108)$$

As an application the pitched roof frame of Figure 10.5 is designed. This frame is made out of 254 mm×101·60 mm×22·25 kg per m length. UB section with a second moment of area I of $28·71 \times 10^6$ mm⁴. An elastic-plastic analysis shows that the ultimate carrying capacity of the frame is satisfactory. It is required to obtain a

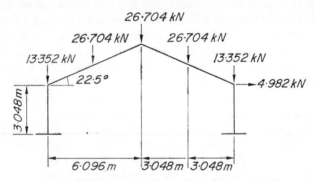

Figure 10.5. Pitched roof frame-dimensions and loading

minimum weight design for the frame that satisfies two deflection restrictions. The elastic maximum sway of the eaves, at the working loads, should not exceed 9·37 mm. This corresponds to 1/325 of the column height. The maximum vertical deflection of the apex should not exceed 37·5 mm and this corresponds to 1/325 of the span.

The support moments M_A and M_E and the internal moments M_D were selected as the redundant forces. Trial analyses showed the upper and lower bounds for the second moments of area I_1 and I_2 for the columns and the rafters to be given by

$$28·71 \times 10^6 \leqslant I_1 \leqslant 156·46 \times 10^6 \ mm^4,$$
$$28·71 \times 10^6 \leqslant I_2 \leqslant 156·46 \times 10^6 \ mm^4.$$

Accordingly the upper and lower bounds of the redundants are
$$36·18 \leqslant M_A \leqslant 93·61 \ kNm$$
$$-59·46 \leqslant M_D \leqslant -21·76 \ kNm$$
$$-107·53 \leqslant M_E \leqslant -44·53 \ kNm$$

The design procedure involved eighty-five reductions of the simplex table to obtain two local optima. The second and the lower of the two gave $I_1 = 44 \cdot 49 \times 10^6$ mm⁴ and $I_2 = 73 \cdot 30 \times 10^6$ mm⁴. The corresponding sway of column DE and the vertical deflection of the apex were 9·37 mm and 20·27 mm respectively. Thus the sway deflection of the column had reached its permissible value. The second moments of area obtained by the design procedure were used in a check analysis of the frame. This showed that the sway of column DE was 9·37 mm and the vertical deflection of the apex was 20·19 mm.

EXERCISES

1. Design the pin jointed frame in Figure 10.6 for minimum weight. The vertical deflection at C is limited to a maximum of 0·254 mm. The maximum tensile stress should not exceed 146·68 N/mm² Use Rankine formula to limit the compressive stress.

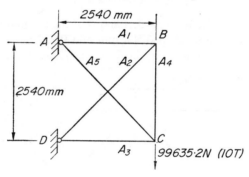

Figure 10.6

There is a lower bound of 1290·32 mm² on the areas. Take $E = 206 \cdot 896$ kN/mm²

Answer.
$$A_1 = A_4 = 1290 \cdot 32 \text{ mm}^2$$
$$A_2 = 2490 \cdot 32 \text{ mm}^2$$
$$A_3 = 10851 \cdot 59 \text{ mm}^2$$
$$A_5 = 21270 \cdot 93 \text{ mm}^2$$

2. The horizontal and vertical deflections at joint B of the pin-jointed frame in Figure 10.7 are limited to a maximum value of 5·08 mm. The maximum tensile and compressive stresses should

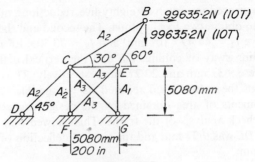

Figure 10.7

not exceed 146·68 N/mm². Design the frame for minimum weight.
Take $E = 206·896$ kN/mm².

Answer. $A_1 = 45909·59$ mm²
                              $A_2 = 42916·04$ mm²
                              $A_3 = 13980·62$ mm²

3. The vertical deflections at A and B of the pin-jointed frame in
Figure 10.8 are limited to 0·508 mm and 0·254 mm respectively.
The cross-sectional areas of the vertical, horizontal and diagonal

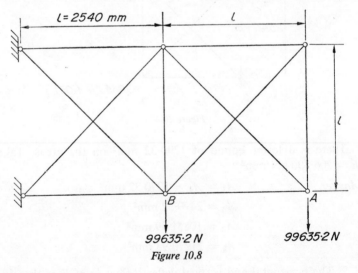

Figure 10.8

members are A_1, A_2 and A_3 respectively. Design the frame for
minimum weight.

314

Answer.
$$A_1 = 18290.29 \text{ mm}^2$$
$$A_2 = 26025.75 \text{ mm}^2$$
$$A_3 = 3225.80 \text{ mm}^2$$

4. It is intended to build a pin jointed frame *BAC* so that it carries a vertical load *V* and a horizontal load *H* at *A* (Figure 10.9). The load *V* always acts downwards while *H* may reverse its direction. The two members are of equal constant length *L*, while the

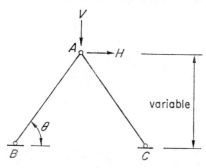

Figure 10.9

distance between *B* and *C* may vary thus varying the angle θ. The horizontal deflection at *A* is limited to a maximum value of 0.25 mm while the maximum compressive stress in the members should not exceed 0.1 kN/mm². For $L = 1000$ mm, $V = \pm H = 10$ kN, design the frame for minimum weight.

Answer. $\theta = 46.5°$, area of section 141 mm².

APPENDIX 1

STABILITY FUNCTIONS FOR COMPRESSIVE FORCES

ϱ	ϕ_1	ϕ_2	ϕ_3	ϕ_4	ϕ_5	S	C
0·00	1·0000	1·0000	1·0000	1·0000	1·0000	4·0000	0·5000
0·02	0·9865	0·9967	0·9934	1·0033	0·9803	3·9736	0·5050
0·04	0·9669	0·9934	0·9868	1·0067	0·9605	3·9471	0·5101
0·06	0·9592	0·9901	0·9801	1·0101	0·9407	3·9204	0·5153
0·08	0·9333	0·9868	0·9734	1·0135	0·9210	3·8936	0·5206
0·10	0·9154	0·9834	0·9667	1·0170	0·9012	3·8667	0·5260
0·12	0·8993	0·9801	0·9599	1·0205	0·8814	3·8396	0·5316
0·14	0·8821	0·9767	0·9531	1·0241	0·8616	3·8123	0·5372
0·16	0·8648	0·9734	0·9462	1·0277	0·8418	3·7849	0·5430
0·18	0·8474	0·9700	0·9394	1·0313	0·8220	3·7574	0·5490
0·20	0·8298	0·9666	0·9324	1·0350	0·8021	3·7297	0·5550
0·22	0·8122	0·9632	0·9255	1·0388	0·7823	3·7019	0·5612
0·24	0·7943	0·9598	0·9185	1·0426	0·7624	3·6739	0·5676
0·26	0·7764	0·9564	0·9114	1·0464	0·7426	3·6457	0·5741
0·28	0·7584	0·9530	0·9043	1·0503	0·7227	3·6174	0·5807
0·30	0·7402	0·9496	0·8972	1·0543	0·7028	3·5889	0·5875
0·32	0·7218	0·9461	0·8901	1·0583	0·6829	3·5602	0·5945
0·34	0·7034	0·9427	0·8828	1·0623	0·6630	3·5314	0·6017
0·36	0·6848	0·9392	0·8756	1·0665	0·6431	3·5024	0·6090
0·38	0·6660	0·9357	0·8683	1·0706	0·6232	3·4732	0·6165
0·40	0·6471	0·9323	0·8610	1·0748	0·6033	3·4439	0·6242

0·6321	3·4144	0·5833	1·0791	0·8536	0·9288	0·6281	0·42
0·6402	3·3847	0·5634	1·0835	0·8462	0·9253	0·6089	0·44
0·6485	3·3548	0·5434	1·0878	0·8387	0·9217	0·5895	0·46
0·6572	3·3247	0·5234	1·0923	0·8312	0·9182	0·5701	0·48
0·6659	3·2945	0·5035	1·0968	0·8236	0·9147	0·5504	0·50
0·6749	3·2640	0·4835	1·1014	0·8160	0·9111	0·5306	0,52
0·6841	3·2334	0·4634	1·1060	0·8083	0·9076	0·5106	0·54
0·6937	3·2025	0·4434	1·1108	0·8006	0·9040	0·4905	0·56
0·7035	3·1715	0·4234	1·1155	0·7929	0·9004	0·4702	0·58
0·7136	3·1403	0·4034	1·1204	0·7851	0·8968	0·4498	0·60
0·7239	3·1088	0·3833	1·1253	0·7772	0·8932	0·4291	0·62
0·7346	3·0771	0·3632	1·1303	0·7693	0·8896	0·4083	0·64
0·7456	3·0453	0·3432	1·1353	0·7613	0·8860	0·3873	0·66
0·7570	3·0132	0·3231	1·1404	0·7533	0·8823	0·3661	0·68
0·7687	2·9809	0·3030	1·1456	0·7452	0·8787	0·3448	0·70
0·7807	2·9484	0·2829	1·1509	0·7371	0·8750	0·3233	0·72
0·7932	2·9156	0·2627	1·1563	0·7289	0·8714	0·3015	0·74
0·8060	2·8826	0·2426	1·1617	0·7207	0·8677	0·2796	0·76
0·8193	2·8494	0·2225	1·1672	0·7123	0·8640	0·2575	0·78
0·8330	2·8159	·02023	1·1728	0·7040	0·8603	0·2351	0·80
0·8472	2·7822	0·1821	1·1785	0·6956	0·8565	0·2126	0·82
0·8618	2·7483	0·1619	1·1843	0·6871	0·8528	0·1899	0·84
0·8770	2·7141	0·1417	1·1901	0·6735	0·8491	0·1669	0·86
0·8927	2·6797	0·1215	1·1961	0·6699	0·8453	0·1438	0·88
0·9090	2·6450	0·1013	1·2021	0·6612	0·8415	0·1204	0·90
0·9258	2·6100	0·0811	1·2082	0·6525	0·8377	0·0968	0·92
0·9433	2·5748	0·0608	1·2144	0·6437	0·8339	0·0729	0·94
0·9615	2·5392	0·0406	1·2208	0·6348	0·8301	0·0489	0·96
0·9804	2·5035	0·0203	1·2272	0·6259	0·8263	0·0246	0·98
1·0000	2·4674	—0·0000	1·2337	0·6169	0·8225	—0·0000	1·00

ϱ	ϕ_1	ϕ_2	ϕ_3	ϕ_4	ϕ_5	S	C
1·02	—0·0248	0·8186	0·6078	1·2403	—0·0203	2·4311	1·0204
1·04	—0·0498	0·8148	0·5986	1·2471	—0·0406	2·3944	1·0416
1·06	—0·0752	0·8109	0·5894	1·2539	—0·0609	2·3575	1·0638
1·08	—0·1007	0·8070	0·5801	1·2608	—0·0813	2·3202	1·0868
1·10	—0·1266	0·8031	0·5707	1·2679	—0·1016	2·2827	1·1109
1·12	—0·1527	0·7992	0·5612	1·2751	—0·1220	2·2448	1·1360
1·14	—0·1790	0·7952	0·5517	1·2824	—0·1424	2·2066	1·1623
1·16	—0·2057	0·7913	0·5420	1·2898	—0·1628	2·1681	1·1898
1·18	—0·2327	0·7873	0·5323	1·2973	—0·1832	2·1293	1·2185
1·20	—0·2599	0·7833	0·5225	1·3050	—0·2036	2·0901	1·2487
1·22	—0·2875	0·7794	0·5126	1·3128	—0·2241	2·0506	1·2804
1·24	—0·3153	0·7754	0·5027	1·3207	—0·2445	2·0107	1·3137
1·26	—0·3435	0·7713	0·4926	1·3288	—0·2650	1·9705	1·3487
1·28	—0·3720	0·7673	0·4825	1·3370	—0·2855	1·9299	1·3855
1·30	—0·4009	0·7633	0·4722	1·3453	—0·3060	1·8889	1·4244
1·32	—0·4300	0·7592	0·4619	1·3538	—0·3265	1·8476	1·4655
1·34	—0·4595	0·7551	0·4515	1·3624	—0·3470	1·8058	1·5089
1·36	—0·4894	0·7510	0·4409	1·3712	—0·3675	1·7637	1·5549
1·38	—0·5196	0·7469	0·4303	1·3802	—0·3881	1·7212	1·6038
1·40	—0·5502	0·7428	0·4196	1·3893	—0·4087	1·6782	1·6557
1·42	—0·5811	0·7387	0·4087	1·3985	—0·4292	1·6348	1·7109
1·44	—0·6125	0·7345	0·3978	1·4080	—0·4499	1·5910	1·7699
1·46	—0·6442	0·7303	0·3867	1·4176	—0·4705	1·5468	1·8329
1·48	—0·6763	0·7261	0·3755	1·4247	—0·4911	1·5021	1·9005
1·50	—0·7089	0·7219	0·3642	1·4373	—0·5118	1·4570	1·9731

1·52	2·0512	1·4114	—0·5324	1·4475	0·3528	0·7177	—0·7418
1·54	2·1356	1·3653	—0·5531	1·4578	0·3413	0·7135	—0·7752
1·56	2·2271	1·3187	—0·5738	1·4684	0·3297	0·7092	—0·8090
1·58	2·3263	1·2716	—0·5945	1·4791	0·3179	0·7050	—0·8433
1·60	2·4348	1·2240	—0·6153	1·4901	0·3060	0·7007	—0·8781
1·62	2·5534	1·1759	—0·6360	1·5012	0·2940	0·6964	—0·9133
1·64	2·6838	1·1272	—0·6568	1·5126	0·2818	0·6921	—0·9490
1·66	2·8278	1·0780	—0·6776	1·5242	0·2695	0·6877	—0·9852
1·68	2·9877	1·0282	—0·6984	1·5360	0·2571	0·6834	—1·0219
1·70	3·1662	0·9779	—0·7192	1·5481	0·2445	0·6790	—1·0592
1·72	3·3667	0·9270	—0·7400	1·5604	0·2317	0·6746	—1·0969
1·74	3·5936	0·8754	—0·7609	1·5730	0·2189	0·6702	—1·1353
1·76	3·8524	0·8233	—0·7817	1·5858	0·2058	0·6658	—1·1741
1·78	4·1504	0·7705	—0·8026	1·5988	0·1926	0·6614	—1·2136
1·80	4·4969	0·7170	—0·8235	1·6122	0·1793	0·6569	—1·2537
1·82	4·9051	0·6629	—0·8445	1·6258	0·1657	0·6524	—1·2944
1·84	5·3929	0·6081	—0·8654	1·6397	0·1520	0·6479	—1·3357
1·86	5·9859	0·5526	—0·8864	1·6539	0·1382	0·6434	—1·3776
1·88	6·7223	0·4964	—0·9074	1·6684	0·1241	0·6389	—1·4202
1·90	7·6612	0·4394	—0·9284	1·6833	0·1099	0·6343	—1·4635
1·92	8·8990	0·3817	—0·9494	1·6984	0·0954	0·6298	—1·5076
1·94	10·6056	0·3232	—0·9704	1·7139	0·0808	0·6252	—1·5523
1·96	13·1087	0·2639	—0·9915	1·7297	0·0660	0·6206	—1·5978
1·98	17·1355	0·2038	—1·0126	1·7459	0·0509	0·6159	—1·6440
2·00	24·6841	0·1428	—1·0337	1·7624	0·0357	0·6113	—1·6910
2·02	43·9616	0·0809	—1·0548	1·7793	0·0202	9·6066	—1·8388
2·04	197·3864	0·0182	—1·0759	1·7966	0·0046	0·6019	—1·7875

ϱ	ϕ_1	ϕ_2	ϕ_3	ϕ_4	ϕ_5	S	C
2·06	—1·8371	0·5972	—0·0114	1·8143	—1·0971	—0·0455	—79·8138
2·08	—1·8875	0·5925	—0·0275	1·8324	—1·1183	—0·1101	—33·2921
2·10	—1·9388	0·5877	—0·0439	1·8510	—1·1395	—0·1757	—21·0722
2·12	—1·9911	0·5829	—0·0606	1·8700	—1·1607	—0·2423	—15·4361
2·14	—2·0444	0·5781	—0·0775	1·8894	—1·1819	—0·3099	—12·1925
2·16	—2·0986	0·5733	—0·0947	1·9093	—1·2032	—0·3786	—10·0850
2·18	—2·1539	0·5685	—0·1121	1·9297	—1·2245	—0·4485	—8·6059
2·20	—2·2163	0·5636	—0·1299	1·9506	—1·2458	—0·5194	—7·5107
2·22	—2·2678	0·5587	—0·1479	1·9720	—1·2671	—0·5916	—6·6673
2·24	—2·3264	0·5538	—0·1662	1·9940	—1·2885	—0·6649	—5·9978
2·26	—2·3863	0·5489	—0·1849	2·0165	—1·3099	—0·7395	—5·4537
2·28	—2·4473	0·5440	—0·2039	2·0396	—1·3313	—0·8154	—5·0027
2·30	—2·5096	0·5390	—0·2232	2·0633	—1·3527	—0·8926	—4·6230
2·32	—2·5733	0·5340	—0·2428	2·0876	—1·3741	—0·9713	—4·2988
2·34	—2·6383	0·5290	—0·2628	2·1126	—1·3956	—1·0513	—4·0190
2·36	—2·7047	0·5239	—0·2832	2·1382	—1·4171	—1·1328	—3·7750
2·38	—2·7725	0·5189	—0·3040	2·1646	—1·4386	—1·2159	—3·5604
2·40	—2·8419	0·5138	—0·3251	2·1916	—1·4601	—1·3006	—3·3703
2·42	—2·9129	0·5087	—0·3467	2·2195	—1·4817	—1·3869	—3·2006
2·44	—2·9855	0·5035	—0·3687	2·2480	—1·5033	—1·4749	—3·0484
2·46	—3·0598	0·4984	—0·3912	2·2774	—1·5249	—1·5647	—2·9111
2·48	—3·1359	0·4932	—0·4141	2·3077	—1·5465	—1·6563	—2·7865
2·50	—3·2138	0·4880	—0·4375	2·3388	—1·5682	—1·7499	—2·6732

2·52	—3·2936	0·4827	—0·4613	2·3709	—1·5899	—1·8454	—2·5695
2·54	—3·3754	0·4775	—0·4857	2·4039	—1·6116	—1·9430	—2·4744
2·56	—3·4592	0·4722	—0·5107	2·4379	—1·6333	—2·0427	—2·3869
2·58	—3·5452	0·4669	—0·5362	2·4729	—1·6551	—2·1447	—2·3061
2·60	—3·6335	0·4615	—0·5622	2·5090	—1·6769	—2·2490	—2·2312
2·62	—3·7241	0·4561	—0·5889	2·5463	—1·6987	—2·3557	—2·1618
2·64	—3·8172	0·4507	—0·6162	2·5874	—1·7206	—2·4650	—2·0971
2·66	—3·9128	0·4453	—0·6442	2·6244	—1·7424	—2·5769	—2·0369
2·68	—4·0111	0·4399	—0·6729	2·6654	—1·7643	—2·6915	—1·9805
2·70	—4·1122	0·4344	—0·7023	2·7077	—1·7863	—2·8091.	—1·9278
2·72	—4·2162	0·4289	—0·7324	2·7514	—1·8082	—2·9296	—1·8784
2·74	—4·3233	0·4233	—0·7633	2·7967	—1·8302	—3·0533	—1·8319
2·76	—4·4336	0·4178	—0·7951	2·8435	—1·8522	—3·1803	—1·7882
2·78	—4·5473	0·4122	—0·8277	2·8919	—1·8743	—3·3108	—1·7470
2·80	—4·6645	0·4066	—0·8612	2·9421	—1·8964	—3·4449	—1·7081
2·82	—4·7854	0·4009	—0·8957	2·9941	—1·9185	—3·5828	—1·6714
2·84	—4·9103	0·3952	—0·9312	3·0480	—1·9406	—3·7246	—1·6366
2·86	—5·0392	0·3895	—0·9677	3·1039	—1·9628	—3·8707	—1·6038
2·88	—5·1725	0·3837	—1·0053	3·1619	—1·9850	—4·0213	—1·5726
2·90	—5·3104	0·3780	—1·0441	3·2222	—2·0072	—4·1765	—1·5430
2·92	—5·4531	0·3722	—1·0841	3·2848	—2·0294	—4·3306	—1·5149
2·94	—5·6009	0·3663	—1·1255	3·3499	—2·0517	—4·5019	—1·4882
2·96	—5·7540	0·3605	—1·1682	3·4177	—2·0740	—4·6727	—1·4628
2·98	—5·9129	0·3545	—1·2123	3·4883	—2·0964	—4·8492	—1·4387
3·00	—6·0778	0·3486	—1·2580	3·5618	—2·1188	—5·0320	—1·4157
3·02	—6·2491	0·3426	—1·3053	3·6358	—2·1412	—5·2212	—1·3937
3·04	—6·4273	0·3366	—1·3543	3·7186	—2·1637	—5·4174	—1·3728

22

ϱ	ϕ_1	ϕ_2	ϕ_3	ϕ_4	ϕ_5	S	C
3·06	—6·6127	0·3306	—1·4052	3·8023	—2·1862	—5·6209	—1·3529
3·08	—6·8058	0·3245	—1·4581	3·8897	—2·2087	—5·8223	—1·3339
3·10	—7·0072	0·3184	—1·5130	3·9812	—2·2312	—6·0519	—1·3157
3·12	—7·2174	0·3123	—1·5701	4·0771	—2·2538	—6·2805	—1·2983
3·14	—7·4369	0·3061	—1·6297	4·1776	—2·2764	—6·5186	—1·2817
3·16	—7·6666	0·2999	—1·6917	4·2831	—2·2991	—6·7669	—1·2659
3·18	—7·9071	0·2936	—1·7565	4·3940	—2·3218	—7·0262	—1·2508
3·20	—8·1592	0·2874	—1·8243	4·5106	—2·3445	—7·2971	—1·2363
3·22	—8·4238	0·2810	—1·8952	4·6334	—2·3673	—7·5807	—1·2224
3·24	—8·7019	0·2747	—1·9695	4·7630	—2·3901	—7·8779	—1·2092
3·26	—8·9947	0·2683	—2·0475	4·8997	—2·4130	—8·1899	—1·1965
3·28	—9·3032	0·2618	—2·1294	5·0444	—2·4359	—8·5178	—1·1844
3·30	—9·6290	0·2554	—2·2157	5·1975	—2·4588	—8·8629	—1·1729
3·32	—9·9734	0·2488	—2·3067	5·3600	—2·4818	—9·2269	—1·1618
3·34	—10·3382	0·2423	—2·4028	5·5325	—2·5048	—9·6114	—1·1512
3·36	—10·7253	0·2357	—2·5046	5·7162	—2·5278	—10·0183	—1·1412
3·38	—11·1369	0·2290	—2·6124	5·9120	—2·5509	—10·4497	—1·1315
3·40	—11·5754	0·2224	—2·7271	6·1212	—2·5740	—10·9082	—1·1223
3·42	—12·0435	0·2157	—2·8491	6·3452	—2·5972	—11·3965	—1·1135
3·44	—12·5445	0·2089	—2·9795	6·5856	—2·6204	—11·9178	—1·1052
3·46	—13·0820	0·2021	—3·1189	6·8441	—2·6437	—12·4757	—1·0972
3·48	—13·6602	0·1952	—3·2686	7·1230	—2·6670	—13·0745	—1·0896
3·50	—14·2840	0·1883	—3·4297	7·4245	—2·6903	—13·7190	—1·0824

3·52	—14·9591	0·1814	—3·6037	7·7517	—2·7137	—14·4149	—1·0755
3·54	—15·6922	0·1744	—3·7922	8·1077	—2·7371	—15·1689	—1·0690
3·56	—16·4912	0·1674	—3·9973	8·4967	—2·7606	—15·9890	—1·0628
3·58	—17·3655	0·1603	—4·2211	8·9232	—2·7841	—16·8845	—1·0570
3·60	—18·3264	0·1532	—4·4667	9·3930	—2·8077	—17·8668	—1·0514
3·62	—19·3875	0·1460	—4·7374	9·9128	—2·8313	—18·9494	—1·0462
3·64	—20·5657	0·1388	—5·0373	10·4911	—2·8550	—20·1492	—1·0413
3·66	—21·8814	0·1316	—5·3717	11·1380	—2·8787	—21·4868	—1·0367
3·68	—23·3606	0·1242	—5·7470	11·8667	—2·9024	—22·9879	—1·0324
3·70	—25·0358	0·1169	—6·1713	12·6932	—2·9262	—24·6852	—1·0284
3·72	—26·9492	0·1095	—6·6552	13·6388	—2·9501	—26·6208	—1·0247
3·74	—29·1556	0·1020	—7·2124	14·7308	—2·9740	—28·8496	—1·0212
3·76	—31·7284	0·0945	—7·8612	16·0059	—2·9980	—31·4449	—1·0180
3·78	—34·7673	0·0869	—8·6266	17·5140	—3·0220	—34·5066	—1·0151
3·80	—38·4124	0·0793	—9·5436	19·3251	—3·0461	—38·1745	—1·0125
3·82	—42·3655	0·0716	—10·6627	21·5402	—3·0702	—42·0506	—1·1001
3·84	—48·4298	0·0639	—12·0595	24·3107	—3·0944	—48·2381	—1·0079
3·86	—55·5813	0·0561	—13·8533	27·8748	—3·1186	—55·4130	—1·0061
3·88	—65·1139	0·0483	—16·2423	32·6293	—3·1429	—64·9691	—1·0045
3·90	—78·4560	0·0404	—19·5837	39·2885	—3·1673	—78·3349	—1·0031
3·92	—98·4647	0·0324	—24·5919	49·2810	—3·1917	—98·3675	—1·0020
3·94	—131·8069	0·0244	—32·9334	65·9400	—3·2161	—131·7337	—1·0101
3·96	—198·4823	0·0163	—49·6083	99·2557	—3·2406	—198·4334	—1·0005
3·98	—398·4912	0·0082	—99·6166	199·2579	—3·2652	—398·4666	—1·0001

STABILITY FUNCTIONS FOR TENSILE FORCES

ϱ	ϕ_1	ϕ_2	ϕ_3	ϕ_4	ϕ_5	S	C
0·00	1·0000	1·0000	1·0000	1·0000	1·0000	4·0000	0·5000
—0·02	1·0164	1·0033	1·0066	0·9967	1·0197	4·0263	0·4951
—0·04	1·0327	1·0066	1·0131	0·9935	1·0395	4·0524	0·4903
—0·06	1·0489	1·0098	1·0196	0·9903	1·0592	4·0784	0·4856
—0·08	1·0649	1·0131	1·0261	0·9872	1·0789	4·1042	0·4810
—0·10	1·0809	1·0163	1·0325	0·9840	1·0986	4·1299	0·4765
—0·12	1·0968	1·0196	1·0389	0·9810	1·1183	4·1555	0·4721
—0·14	1·1126	1·0228	1·0452	0·9779	1·1380	4·1810	0·4678
—0·16	1·1283	1·0260	1·0516	0·9749	1·1576	4·2063	0·4635
—0·18	1·1438	1·0292	1·0579	0·9719	1·1773	4·2316	0·4594
—0·20	1·1593	1·0324	1·0642	0·9690	1·1969	4·2567	0·4553
—0·22	1·1747	1·0356	1·0704	0·9661	1·2166	4·2816	0·4513
—0·24	1·1900	1·0388	1·0766	0·9632	1·2362	4·3065	0·4473
—0·26	1·2052	1·0420	1·0828	0·9604	1·2558	4·3312	0·4435
—0·28	1·2203	1·0452	1·0890	0·9576	1·2755	4·3559	0·4397
—0·30	1·2354	1·0483	1·0951	0·9548	1·9521	4·3804	0·4360
—0·32	1·2503	1·0515	1·1012	0·9521	1·3147	4·4048	0·4323
—0·34	1·2652	1·0546	1·1073	0·9494	1·3343	4·4291	0·4287
—0·36	1·2799	1·0578	1·1133	0·9467	1·3539	4·4532	0·4252
—0·38	1·2946	1·0609	1·1193	0·9441	1·3734	4·4773	0·4217
—0·40	1·3092	1·0640	1·1253	0·9414	1·3930	4·5013	0·4183
—0·42	1·3237	1·0671	1·1313	0·9388	1·4126	4·5251	0·4149
—0·44	1·3381	1·0702	1·1372	0·9363	1·4321	4·5488	0·4117

−0·46	1·3525	1·0733	1·1431	0·9338	1·4517	4·5725	0·4085
−0·48	1·3668	1·0764	1·1490	0·9312	1·4712	4·5960	0·4052
−0·50	1·3809	1·0795	1·1549	0·9288	1·4907	4·6194	0·4021
−0·52	1·3951	1·0826	1·1607	0·9263	1·5103	4·6428	0·3990
−0·54	1·4091	1·0856	1·1665	0·9239	1·5298	4·6660	0·3960
−0·56	1·4231	1·0887	1·1723	0·9215	1·5493	4·6891	0·3930
−0·58	1·4369	1·0917	1·1780	0·9191	1·5688	4·7122	0·3901
−0·60	1·4508	1·0948	1·1838	0·9168	1·5883	4·7351	0·3872
−0·62	1·4645	1·0978	1·1895	0·9145	1·6077	4·7579	0·3844
−0·64	1·4782	1·1008	1·1952	0·9122	1·6272	4·7807	0·3816
−0·66	1·4918	1·1039	1·2008	0·9099	1·6467	4·8033	0·3789
−0·68	1·5053	1·1069	1·2065	0·9077	1·6661	4·8259	0·3762
−0·70	1·5187	1·1099	1·2121	0·9054	1·6856	4·8483	0·3735
−0·72	1·5321	1·1129	1·2177	0·9032	1·7050	4·8707	0·3709
−0·74	1·5454	1·1158	1·2232	0·9011	1·7245	4·8930	0·3683
−0·76	1·5587	1·1188	1·2288	0·8989	1·7439	4·9152	0·3658
−0·78	1·5719	1·1218	1·2343	0·8968	1·7633	4·9373	0·3633
−0·80	1·5850	1·1248	1·2398	0·8947	1·7827	4·9593	0·3608
−0·82	1·5980	1·1277	1·2453	0·8926	1·8021	4·9812	0·3584
−0·84	1·6110	1·1307	1·2508	0·8905	1·8215	5·0031	0·3560
−0·86	1·6239	1·1336	1·2562	0·8885	1·8409	5·0248	0·3536
−0·88	1·6368	1·1366	1·2616	0·8864	1·8603	5·0465	0·3513
−0·90	1·6496	1·1395	1·2670	0·8844	1·8797	5·0681	0·3490
−0·92	1·6623	1·1424	1·2724	0·8824	1·8991	5·0896	0·3468
−0·94	1·6750	1·1453	1·2778	0·8805	1·9184	5·1110	0·3445
−0·96	1·6876	1·1482	1·2831	0·8785	1·9378	5·1323	0·3424
−0·98	1·7002	1·1511	1·2884	0·8766	1·9572	5·1536	0·3402
−1·00	1·7127	1·1540	1·2937	0·8747	1·9765	5·1748	0·3381

ϱ	ϕ_1	ϕ_2	ϕ_3	ϕ_4	ϕ_5	S	C
1·02	1·7251	1·1569	1·2990	0·8728	1·9958	5·1959	0·3360
1·04	1·7375	1·1598	1·3042	0·8710	2·0152	5·2169	0·3339
1·06	1·7498	1·1627	1·3095	0·8691	2·0345	5·2379	0·3319
1·08	1·7621	1·1655	1·3147	0·8673	2·0538	5·2587	0·3298
1·10	1·7743	1·1684	1·3199	0·8655	2·0731	5·2795	0·3279
1·12	1·7865	1·1713	1·3251	0·8637	2·0924	5·3003	0·3259
1·14	1·7986	1·1741	1·3302	0·8619	2·1117	5·3209	0·3240
1·16	1·8106	1·1770	1·3354	0·8601	2·1310	5·3415	0·3221
1·18	1·8226	1·1798	1·3405	0·8584	2·1503	5·3620	0·3202
1·20	1·8346	1·1826	1·3456	0·8566	2·1696	5·3824	0·3183
1·22	1·8464	1·1854	1·3507	0·8549	2·1886	5·4028	0·3165
1·24	1·8583	1·1883	1·3558	0·8532	2·2081	5·4231	0·3147
1·26	1·8701	1·1911	1·3608	0·8516	2·2274	5·4433	0·3129
1·28	1·8818	1·1939	1·3659	0·8499	2·2466	5·4634	0·3111
1·30	1·8935	1·1967	1·3709	0·8483	2·2659	5·4835	0·3094
1·32	1·9051	1·1995	1·3759	0·8466	2·2851	5·5035	0·3077
1·34	1·9167	1·2022	1·3809	0·8450	2·3043	5·5234	0·3060
1·36	1·9282	1·2050	1·3858	0·8434	2·3236	5·5433	0·3043
1·38	1·9397	1·2078	1·3908	0·8418	2·3428	5·5631	0·3026
1·40	1·9512	1·2106	1·3957	0·8402	2·3620	5·5828	0·3010
1·42	1·9626	1·2133	1·4006	1·8387	2·3812	5·6025	0·2994
1·44	1·9739	1·2161	1·4055	0·8371	2·4004	5·6221	0·2978
1·46	1·9852	1·2188	1·4104	0·8356	2·4196	5·6417	0·2962
1·48	1·9965	1·2216	1·4153	0·8341	2·4388	5·6611	0·2947
1·50	2·0077	1·2243	1·4201	0·8326	2·4580	5·6806	0·2931

—1·52	2·0188	1·2270	1·4250	0·8311	2·4772	5·6999	0·2916
—1·54	2·0300	1·2297	1·4298	0·8296	2·4963	5·7192	0·2901
—1·56	2·0410	1·2325	1·4346	0·8282	2·5155	5·7384	0·2886
—1·58	2·0521	1·2352	1·4394	0·8267	2·5347	5·7576	0·2872
—1·60	2·0631	1·2379	1·4442	0·8253	2·5538	5·7767	0·2857
—1·62	2·0740	1·2406	1·4489	0·8239	2·5730	5·7958	0·2843
—1·64	2·0849	1·2433	1·4537	0·8225	2·5921	5·8137	0·2829
—1·66	2·0958	1·2460	1·4584	0·8211	2·6113	5·8337	0·2815
—1·68	2·1066	1·2487	1·4631	0·8197	2·6304	5·8525	0·2801
—1·70	2·1174	1·2513	1·4678	0·8183	2·6495	5·8714	0·2787
—1·72	2·1281	1·2540	1·4725	0·8169	2·6686	5·8901	0·2774
—1·74	2·1388	1·2567	1·4772	0·8156	2·6878	5·9088	0·2761
—1·76	2·1495	1·2593	1·4819	0·8143	2·7069	5·9274	0·2747
—1·78	2·1601	1·2620	1·4865	0·8129	2·7260	5·9460	0·2734
—1·80	2·1706	1·2646	1·4911	0·8116	2·7451	5·9645	0·2721
—1·82	2·1812	1·2675	1·4958	0·8103	2·7642	5·9830	0·2709
—1·84	2·1917	1·2699	1·5004	0·9080	2·7833	6·0014	0·2696
—1·86	2·2021	1·2725	1·5049	0·8078	2·8023	6·0198	0·2684
—1·88	2·2126	1·2752	1·5095	0·8065	2·8214	6·0381	0·2671
—1·90	2·2230	1·2778	1·5141	0·8052	2·8405	6·0564	0·2659
—1·92	2·2333	1·2804	1·5186	0·8040	2·8596	6·0745	0·2647
—1·94	2·2436	1·2830	1·5232	0·8027	2·8786	6·0927	0·2635
—1·96	2·2539	1·2856	1·5277	0·8015	2·8977	6·1108	0·2623
—1·98	2·2641	1·2882	1·5322	0·8003	2·9167	6·1288	0·2612
—2·00	2·2743	1·2908	1·5367	0·7991	2·2958	6·1468	0·2600
—2·02	2·2845	1·2934	1·5412	0·7979	2·9548	6·1648	0·2589
—2·04	2·2946	1·2960	1·5457	0·7967	2·9738	6·1826	0·2577

ϱ	ϕ_1	ϕ_2	ϕ_3	ϕ_4	ϕ_5	S	C
—2·06	2·3047	1·2986	1·5501	0·7955	2·9929	6·2005	0·2566
—2·08	2·3148	1·3012	1·5546	0·7944	3·0119	6·2183	0·2555
—2·10	2·3248	1·3037	1·5590	0·7932	3·0309	6·2360	0·2544
—2·12	2·3348	1·3063	1·5634	0·7921	3·0499	6·2537	0·2533
—2·14	2·3447	1·3089	1·5678	0·7909	3·0689	6·2713	0·2522
—2·16	2·3547	1·3114	1·5722	0·7898	3·0879	6·2889	0·2512
—2·18	2·3646	1·3140	1·5766	0·7887	3·1069	6·3065	0·2501
—2·20	2·3744	1·3165	1·5810	0·7876	3·1259	6·3239	0·2491
—2·22	2·3842	1·3191	1·5853	0·7865	3·1449	6·3414	0·2480
—2·24	2·3940	1·3216	1·5897	0·7854	3·1639	6·3588	0·2470
—2·26	2·4038	1·3241	1·5940	0·7843	3·1829	6·3761	0·2460
—2·28	2·4135	1·3266	1·5948	0·7832	3·2019	6·3934	0·2450
—2·30	2·4232	1·3292	1·6027	0·7821	3·2208	6·4107	0·2440
—2·32	2·4329	1·3317	1·6070	0·7811	3·2398	6·4279	0·2430
—2·34	2·4425	1·3342	1·6113	0·7800	3·2588	6·4451	0·2421
—2·36	2·4521	1·3367	1·6155	0·7790	3·2777	6·4622	0·2411
—2·38	2·4617	1·3392	1·6198	0·7780	3·2967	6·4793	0·2401
—2·40	2·4712	1·3417	1·6241	0·7769	3·3156	6·4963	0·2392
—2·42	2·4807	1·3442	1·6283	0·7759	3·3346	6·5133	0·2383
—2·44	2·4902	1·3467	1·6326	0·7749	3·3535	6·5302	0·2373
—2·46	2·4997	1·3492	1·6368	0·7739	3·3724	6·5471	0·2364
—2·48	2·5091	1·3516	1·6410	0·7729	3·3913	6·5640	0·2355
—2·50	2·5185	1·3541	1·6452	0·7719	3·4103	6·5808	0·2346

0·2337	6·5975	3·4292	0·7709	1·6494	1·3566	2·5278	—2·52
0·2328	6·6143	3·4481	0·7700	1·6536	1·3590	2·5372	—2·54
0·2319	6·6310	3·4670	0·7690	1·6577	1·3615	2·5465	—2·56
0·2311	6·6476	3·4859	0·7680	1·6619	1·3639	2·5558	—2·58
0·2302	6·6642	3·5048	0·7671	1·6660	1·3664	2·5650	—2·60
0·2294	6·6808	3·5237	0·7662	1·6702	1·3688	2·5742	—2·62
0·2285	6·6973	3·5426	0·7652	1·6743	1·3713	2·5834	—2·64
0·2277	6·7137	3·5615	0·7643	1·6784	1·3737	2·5926	—2·66
0·2269	6·7302	3·5804	0·7634	1·6825	1·3762	2·6017	—2·68
0·2260	6·7466	3·5992	0·7625	1·6866	1·3786	2·6108	—2·70
0·2252	6·7629	3·6181	0·7616	1·6907	1·3810	2·6199	—2·72
0·2244	6·7792	3·6370	0·7607	1·6948	1·3834	2·6290	—2·74
0·2236	6·7955	3·6559	0·7598	1·6989	1·3858	2·6380	—2·76
0·2228	6·8118	3·6747	0·7589	1·7029	1·3883	2·6470	—2·78
0·2220	6·8280	3·6936	0·7580	1·7070	1·3907	2·6560	—2·80
0·2212	6·8441	3·7124	0·7571	1·7110	1·3931	2·6649	—2·82
0·2205	6·8602	3·7313	0·7563	1·7151	1·3955	2·6739	—2·84
0·2197	6·8763	3·7501	0·7554	1·7191	1·3978	2·6828	—2·86
0·2189	6·8924	3·7689	0·7545	1·7231	1·4002	2·6916	—2·88
0·2182	6·9084	3·7878	0·7537	1·7271	1·4026	2·7005	—2·90
0·2174	6·9243	3·8066	0·7528	1·7311	1·4050	2·7093	—2·92
0·2167	6·9403	3·8254	0·7520	1·7351	1·4074	2·7181	—2·94
0·2160	6·9562	3·8443	0·7512	1·7390	1·4098	2·7269	—2·96
0·2152	6·9720	3·8631	0·7504	1·7430	1·4121	2·7375	—2·98
0·2145	6·9878	3·8819	0·7495	1·7470	1·4145	2·7444	—3·00
0·2138	7·0036	3·9007	0·7487	1·7509	1·4168	2·7531	—3·02
0·2131	7·0194	3·9195	0·7479	1·7548	1·4192	2·7618	—3·04

ϱ	ϕ_1	ϕ_2	ϕ_3	ϕ_4	ϕ_5	S	C
—3·06	2·7704	1·4216	1·7588	0·7471	3·9383	7·0351	0·2124
—3·08	2·7791	1·4239	1·7627	0·7463	3·9571	7·0508	0·2117
—3·10	2·7877	1·4262	1·7666	0·7455	3·9759	7·0664	0·2110
—3·12	2·7963	1·4286	1·7705	0·7447	3·9947	7·0820	0·2103
—3·14	2·8048	1·4309	1·7744	0·7440	4·0135	7·0976	0·2096
—3·16	2·8134	1·4332	1·7783	0·7432	4·0322	7·1131	0·2090
—3·18	2·8219	1·4356	1·7822	0·7424	4·0510	7·1286	0·2083
—3·20	2·8304	1·4379	1·7860	0·7417	4·0698	7·1441	0·2076
—3·22	2·8388	1·4402	1·7899	0·7409	4·0886	7·1595	0·2070
—3·24	2·8473	1·4425	1·7937	0·7402	4·1073	7·1749	0·2063
—3·26	2·8557	1·4448	1·7976	0·7394	4·1261	7·1903	0·2057
—3·28	2·8641	1·4472	1·8014	0·7387	4·1448	7·2056	0·2050
—3·30	2·8725	1·4495	1·8052	0·7379	4·1636	7·2209	0·2044
—3·32	2·8809	1·4518	1·8090	0·7372	4·1824	7·2362	0·2038
—3·34	2·8892	1·4541	1·8128	0·7365	4·2011	7·2514	0·2031
—3·36	2·8975	1·4563	1·8166	0·7358	4·2198	7·2666	0·2025
—3·38	2·9058	1·4586	1·8204	0·7350	4·2386	7·2818	0·2019
—3·40	2·9141	1·4609	1·8242	0·7343	4·2573	7·2969	0·2013
—3·42	2·9224	1·4632	1·8280	0·7336	4·2760	7·3120	0·2007
—3·44	2·9306	1·4655	1·8318	0·7329	4·2948	7·3271	0·2001
—3·46	2·9388	1·4678	1·8355	0·7322	4·3135	7·3421	0·1995
—3·48	2·9470	1·4700	1·8393	0·7315	4·3322	7·3571	0·1989
—3·50	2·9552	1·4723	1·8430	0·7308	4·3509	7·3721	0·1983

—3·52	2·9634	1·4746	1·8468	0·7302	4·3696	7·3870	0·1977
—3·54	2·9715	1·4768	1·8505	0·7295	4·3883	7·4019	0·1971
—3·56	2·9796	1·4791	1·8542	0·7288	4·4071	7·4168	0·1965
—3·58	2·9877	1·4813	1·8579	0·7281	4·4258	7·4317	0·1960
—3·60	2·9958	1·4836	1·8616	0·7275	4·4445	7·4465	0·1954
—3·62	3·0038	1·4858	1·8653	0·7268	4·4631	7·4613	0·1948
—3·64	3·0119	1·4881	1·8690	0·7261	4·4818	7·4760	0·1943
—3·66	3·0199	1·4903	1·8727	0·7255	4·5005	7·4908	0·1937
—3·68	3·0279	1·4925	1·8764	0·7248	4·5192	7·5055	0·1932
—3·70	3·0359	1·4948	1·8800	0·7242	4·5379	7·5201	0·1926
—3·72	3·0438	1·4970	1·8837	0·7236	4·5566	7·5348	0·1921
—3·74	3·0518	1·4992	1·8873	0·7229	4·5752	7·5494	0·1915
—3·76	3·0597	1·5014	1·8910	0·7223	4·5930	7·5640	0·1910
—3·78	3·0676	1·5036	1·8946	0·7217	4·6126	7·5785	0·1904
—3·80	3·0755	1·5059	1·8983	1·7210	4·6312	7·5930	0·1899
—3·82	3·0834	1·5081	1·9019	0·7204	4·6499	7·6075	0·1894
—3·84	3·0912	1·5103	1·9055	0·7198	4·6685	7·6220	0·1889
—3·86	3·0990	1·5125	1·9091	0·7192	4·6872	7·6364	0·1884
—3·88	3·1068	1·5147	1·9127	0·7186	4·7058	7·6509	0·1878
—3·90	3·1146	1·5169	1·9163	0·7180	4·7245	7·6652	0·1873
—3·92	3·1224	1·5191	1·9199	0·7174	4·7431	7·6796	0·1868
—3·94	3·1302	1·5212	1·9235	0·7168	4·7618	7·6939	0·1863
—3·96	3·1379	1·5234	1·9271	0·7162	4·7804	7·7082	0·1858
—3·98	3·1456	1·5256	1·9306	0·7156	4·7990	7·7225	0·1853

References

1. HORNE M. R., 'Elastic-Plastic failure loads of Plane frames', *Proc. Royal Soc.*, A, **274** 343–364 (1963).
2. LIVESLEY R. K., 'The application of an eletron Digital Computer to some problems of structural analysis', *Struct. Engr.* **34**, 1–12 (1956).
3. BERRY, A., 'The calculation of stresses in aeroplane spars', *Trans. Royal Aero. Soc.* No. 1 (1916).
4. HORNE M. R., 'The effect of finite deformations in the elastic stability of plane frames', *Proc. Royal Soc.* A, **266.** (1962).
5. JENNINGS A., 'A direct iteration method of obtaining latent roots and vectors of a symmetric matrix', *Proc. Cambridge Philosophical Soc.* vol. 63 (1967).
6. MERCHANT W. *et. al.*, 'Critical loads of tall building frames', *Struct. Engrs.* Vols. 33 and 34 (1956).
7. MAJID K. I., 'An evaluation of the elastic critical load and the Rankine load of frames. *Proc. I.C.E.* **36** 576–593 (March 1967).
8. MONTAGUE P. and WARDLEY C. J. M., 'Elastic critical load of rigidly jointed trusses', *The Engineer*, (Nov. 27, 1964).
9. HEYMAN J., 'An approach to the design of tall buildings', *Proc. I.C.E.* **17** 431 (1960).
10. HORNE M. R., 'The plastic properties of rolled sections', *Brit. Weld. Res. Assn.* (1959).
11. DAVIES J. M., 'Shakedown and incremental collapse in the presence of instability,' *Ph. D. thesis, Manchester University*, (1965).
12. MAJID K. I. and ANDERSON D., 'The computer analysis of large multistorey framed structures', *The Structural Engineer*, No. 11, **46** 357–365 (Nov. 1968).
13. JENNINGS A., 'A compact storage for the solution of symmetric linear simultaneous equations', *The Computer Journal*, **9**, No. 3, (Nov. 1966).
14. MAJID K. I., 'The evaluation of the failure load of plane frames', *Proc. Royal Soc.* A, **306**, 297–311 (1968).
15. MERCHANT W., 'The failure load of rigid jointed frameworks as influenced by stability', *The Structural Engineer* **32** (1954).
16. ARIARATNAM S. T., 'The collapse load of elastic-plastic structures'. *Ph. D. dissertation, Cambridge University*, (1959).
17. *Civil Engineering Research Council*, 'Research and development report', *Proc. I.C.E.* **39**, 477 (1968).

18. HORNE M. R., 'Safe loads on I. Section columns in structures designed by plastic theory', *Proc. I.C.E.* **29**, 137 (Sept. 1964).
19. MAJID K. I., and ANDERSON D., Elastic-plastic design of sway frames by computer,' *Proc. I.C.E.* **41** 705–729 (Dec. 1968).
20. PIPPARD and BAKER, *The analysis of engineering structures*, Edward Arnold
21. BELA KREKO., *Linear programming*, Pitman & Sons Ltd (1968).
22. LLEWELLYN R. W., *Linear programming*, Holt, Rinehart and Winston Inc. (1966).
23. HADLEY G., *Non-linear and dynamic programming*, Addison-Wesley Publishing Co. (1964).
24. TOAKLEY A. R., 'The optimum design of triangulated frameworks', *Research report No. 3*, Department of Civil Engineering, The University of Manchester (1967).

Index